PHYSICAL SCIENCE

ABOUT THE AUTHOR

Richard P. Halpern is an associate professor at Bergen Community College, where he has taught mathematics and the physical sciences for the past twelve years. A graduate of the Bronx High School of Science, Dr. Halpern holds a B.S. degree from City College of City University of New York, an M.S. in physics from the University of Massachusetts, and a Ph.D. in meteorology from New York University. He is a member of Sigma Xi and the New York Academy of Sciences.

PHYSICAL SCIENCE

Richard P. Halpern

BARNES & NOBLE BOOKS
A DIVISION OF HARPER & ROW, PUBLISHERS
New York, Cambridge, Philadelphia
San Francisco, London, Mexico City
São Paulo, Sydney

PHYSICAL SCIENCE. Copyright © 1983 by Richard P. Halpern. All rights reserved. Printed in the United States of America. No part of this book may be used or reproduced in any manner whatsoever without written permission except in the case of brief quotations embodied in critical articles and reviews. For information address Harper & Row, Publishers, Inc., 10 East 53rd Street, New York, N.Y. 10022. Published simultaneously in Canada by Fitzhenry & Whiteside Limited, Toronto.

FIRST EDITION

Designer: C. Linda Dingler

Library of Congress Cataloging in Publication Data

Halpern, Richard P.
    Physical science

    Includes index.
    1. Science. I. Title.
Q158.5.H34 1983   500   82-48100
ISBN 0-06-460195-1 (pbk.)

83 84 85 86 10 9 8 7 6 5 4 3 2 1

To my parents
Irving and Margaret Halpern

Contents

Preface xv

PART ONE: PHYSICS

1 Motion 3

   Position 3
   Linear Motion 4
   Newton's Laws 7
   Two-Dimensional Motion 12
   Uniform Circular Motion 13

2 Momentum, Work, and Energy 21

   Momentum 21
   Work and Power 23
   Energy 24
   Transformation of Energy 27

3 Bulk Matter 30

   States of Matter 30
   Density 31
   Pressure 32
   Fluid Behavior 34
   Bernoulli's Principle 37

4 Kinetic Theory 39

   Internal Energy 39
   Temperature and Heat 39
   Ideal Gases 44
   The Laws of Thermodynamics 47

5 Electricity and Magnetism 50

> Static Electricity 50
> Electric Current 53
> Magnetism 56

6 Wave Motion 67

> Wave Types 67
> Wave Characteristics 70
> Examples of Waves 71
> Wave Behavior 72
> Light 78
> Optics 81

7 Quantum Theory and Relativity 90

> The Dawn of the Quantum Theory 90
> Relativity 94
> Is Classical Physics Wrong? 98

PART TWO: Chemistry

8 The Atom 103

> Thompson's Contributions 103
> Rutherford's Contributions 104
> Atomic Mass and Isotopes 106
> Bohr's Contributions 107
> The Quantum Mechanical Model of the Atom 109

9 Elements and Compounds 114

> Elements 114
> Compounds 117
> Formula Prediction 121
> Polyatomic Ions 122
> The Periodic Table and Electronic Structure 123
> Naming Simple Compounds 123

10 Chemical Equations 125

> Balancing Equations 125
> Types of Equations 126
> Moles 128
> Energy and Chemical Reactions 128

CONTENTS

11 Solutions 132

    Mechanisms of Solution 132
    Describing Solutions 133
    Factors Affecting Solubility 134
    Ionic Equations 134
    Acids and Bases 135
    Chemical Equilibrium 137

12 Electrochemistry 140

    Oxidation-Reduction Reactions 140
    Electrolysis 142
    Voltaic Cells 144

13 Organic Chemistry and Biochemistry 148

    Organic Chemistry 148
    Biochemistry 154

14 The Nucleus 162

    Radioactivity 162
    Nuclear Energy 165

PART THREE: Astronomy

15 Earth, Sun, and Moon 171

    The Earth in Orbit 171
    The Sun 179
    The Moon 180

16 The Solar System 189

    Planetary Motion and Kepler's Laws 189
    The Celestial Sphere 191
    The Planets 193
    Other Solar Bodies 197

17 The Universe 199

    Stars 199
    Galaxies 204
    Cosmology 206

PART FOUR: Meteorology

18 The Atmosphere 211

    Composition of the Atmosphere 211
    Vertical Structure of the Atmosphere 213
    Atmospheric Heating 217

19 Atmospheric Motion 220

    Horizontal Forces 220
    Classification of Winds 223
    Thermally Driven Circulation 225
    General Circulation 228
    Vertical Motion 232

20 Weather Disturbances 234

    Clouds and Precipitation 234
    The Mid-Latitude Cyclone 237

PART FIVE: Geology

21 Introduction to Geology 247

    Minerals, Rocks, and Soils 247
    Streams 250
    Ice and Water 251
    Erosion, Weathering, and Mass Wasting 254

22 The Earth's Interior 260

    Earthquakes 260
    Igneous Activity 263
    Mountains 267
    Groundwater 270

23 Evolution of the Earth 273

    Plate Tectonics 273
    Geologic Dating 277
    Geologic Time 280

24 Pollution 282

    Air Pollution 282
    Water Pollution 284

CONTENTS

Appendix 287

> Signed Numbers 287
> Exponents 288
> Powers of Ten and Scientific Notation 289
> Measurement 291
> Simple Equations 296
> Formula Evaluation 297
> Algebraic Fractions 297
> Graph Interpretation 298
> Scalars and Vectors 298
> Direct and Inverse Proportion 302
> Some Facts About Circles, Parabolas, and Ellipses 303
> Temperature Conversions 304

Index 305

Preface

This book provides a broad overview of the fundamentals of physical science. The approach is relatively nonmathematical; what little math is needed is reviewed in the appendix.

By physical science, we usually mean physics, chemistry, and the earth sciences. This outline is divided into five sections: physics, chemistry, meteorology, astronomy, and geology. These distinctions are for convenience only; particular topics in physical science are not the exclusive property of one type of scientist. Hence, physicists are extremely interested in atoms; meteorologists study the relation between weather and sunspots; geologists look to astronomy for clues about earth history.

For the lay reader with no background in physical science, this book should serve as a introductory text. (Suggestion: read the sections on physics and chemistry first.) Students taking a physical science course will find this outline both a handy adjunct to their present text and an excellent source for review. Instructors contemplating using this as a text should supplement it with readings and questions.

The author would like to thank Jeanne Flagg and the staff of Harper & Row for their help in transforming the original manuscript into a healthy finished product.

PERIODIC TABLE OF THE ELEMENTS

IA	IIA	IIIB	IVB	VB	VIB	VIIB	VIII			IB	IIB	IIIA	IVA	VA	VIA	VIIA	O
1 H 1.00797																	2 He 4.0026
3 Li 6.939	4 Be 9.0122											5 B 10.811	6 C 12.01115	7 N 14.0067	8 O 15.9994	9 F 18.9984	10 Ne 20.183
11 Na 22.9898	12 Mg 24.312											13 Al 26.9815	14 Si 28.086	15 P 30.9738	16 S 32.064	17 Cl 35.453	18 Ar 39.948
19 K 39.102	20 Ca 40.08	21 Sc 44.956	22 Ti 47.90	23 V 50.942	24 Cr 51.996	25 Mn 54.9380	26 Fe 55.847	27 Co 58.9332	28 Ni 58.71	29 Cu 63.54	30 Zn 65.37	31 Ga 69.72	32 Ge 72.59	33 As 74.9216	34 Se 78.96	35 Br 79.909	36 Kr 83.80
37 Rb 85.47	38 Sr 87.62	39 Y 88.905	40 Zr 91.22	41 Nb 92.906	42 Mo 95.94	43 Tc (99)	44 Ru 101.07	45 Rh 102.905	46 Pd 106.4	47 Ag 107.870	48 Cd 112.40	49 In 114.82	50 Sn 118.69	51 Sb 121.75	52 Te 127.60	53 I 126.9044	54 Xe 131.30
55 Cs 132.905	56 Ba 137.34	57 *La 138.91	72 Hf 178.49	73 Ta 180.948	74 W 183.85	75 Re 186.2	76 Os 190.2	77 Ir 192.2	78 Pt 195.09	79 Au 196.967	80 Hg 200.59	81 Tl 204.37	82 Pb 207.19	83 Bi 208.980	84 Po (210)	85 At (210)	86 Rn (222)
87 Fr (223)	88 Ra (226)	89 †Ac (227)															

*Lanthanum Series

58 Ce 140.12	59 Pr 140.907	60 Nd 144.24	61 Pm (145)	62 Sm 150.35	63 Eu 151.96	64 Gd 157.25	65 Tb 158.924	66 Dy 162.50	67 Ho 164.930	68 Er 167.26	69 Tm 168.934	70 Yb 173.04	71 Lu 174.97

†Actinium Series

90 Th 232.038	91 Pa (231)	92 U 238.03	93 Np (237)	94 Pu (242)	95 Am (243)	96 Cm (247)	97 Bk (249)	98 Cf (251)	99 Es (254)	100 Fm (253)	101 Md (256)	102 No (253)	103 Lw (257)

The numbers in parentheses are the mass numbers of most stable or most common isotope.

PART ONE
PHYSICS

1
Motion

The term *motion* implies a change in an object's position in a given time interval. Motion in the real world is complicated. However, we can learn a great deal from the simple cases discussed in this chapter. Here we will consider several aspects of motion and investigate laws that Newton formulated to describe it.

POSITION

How do we measure position? First, we need a point in space which is fixed during the object's motion. Then we imagine an x-y coordinate system to be centered at that point. Such a coordinate system is called a *frame of reference*. Any point firmly attached to the earth's surface can be considered the origin of a frame of reference. Finally we denote the position by specifying the x- and y-coordinates of the object. This is illustrated in figure 1-1. Sometimes an object moves in a circle. In this special case, we indicate position by the angle θ in figure 1-1.

Changes of position, which are vital in the description of motion, can be denoted by specifying one of three possible quantities:

1) The *distance, d,* that an object travels.
2) The *displacement, D,* which is a vector quantity* whose magnitude is the straight line distance between initial and final positions of an object, and whose direction is the direction from the initial to the final position. Note that displacement is not the same thing as distance. If an object moves from some point A to a point B, a

---
*Vector quantities are printed in boldface type.

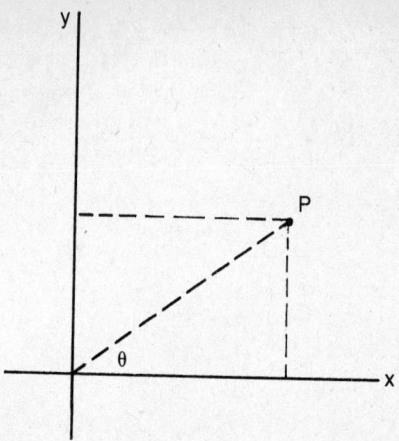

Fig. 1–1 Locating a point P in an $x - y$ coordinate system

distance of 3 meters, and then back again, the displacement is zero but the distance traveles is 6 meters.
3) The *angular displacement*, $\Delta\theta$, which is the change in the angle in the special case of circular motion.

LINEAR MOTION

Speed

1) The *average* speed*, $\bar{v}$, over an interval, is the scalar defined as the distance traveled, d, divided by the elapsed time, t:

$$\bar{v} = d/t$$

The SI units are m/s; the cgs units are cm/s.

Example: An object moves 10 meters in 2 seconds. What is its average speed?
Solution: Using the definition of average speed, we can write:

$$\bar{v} = 10 \text{ m}/2\text{s} = 5 \text{ m/s}$$

---
*Average quantities are indicated by an overbar.

MOTION

2) The *instantaneous speed* is the speed at a particular instant. It is the quantity that is indicated, for example, by a car's speedometer. Unfortunately, expressing it in a neat formula requires calculus. For our purposes, we can regard the instantaneous speed at a time t_0 as the average speed over a very short time interval which begins just before t_0 and ends just afterward.

Velocity

1) The *average velocity*, $\bar{\mathbf{v}}$, over an interval is the displacement divided by the elapsed time:

$$\bar{\mathbf{v}} = \mathbf{D}/t$$

Note that this is a vector, and is in the direction of the displacement. It has the same units as speed.

2) The *instantaneous velocity*, $\mathbf{v}$, is a vector whose magnitude is the instantaneous speed and whose direction is the direction of motion at that instant. A concise expression for this also requires calculus, but the relation between these two types of velocity is analogous to the relation between instantaneous and average speed.

Acceleration

1) The *average acceleration*, $\bar{\mathbf{a}}$, over an interval is the change in velocity divided by the elapsed time:

$$\bar{\mathbf{a}} = \frac{\mathbf{v}_f - \mathbf{v}_i}{t} = \frac{\Delta \mathbf{v}}{\Delta t}$$

where $\mathbf{v}_f$ = final velocity
  $\mathbf{v}_i$ = initial velocity
  t = time
  Δ means "change in"

The SI units are meters per second per second, or m/s^2. The cgs units are cm/s^2. A positive value of $\bar{\mathbf{a}}$ means that the final speed in the interval is greater than the initial speed. Thus, positive acceleration implies increasing speed. A negative value of $\overline{\mathbf{a}}$ means that the final speed for the interval is less than the initial speed. Hence, negative acceleration implies a decreasing speed. This is sometimes called *deceleration*.

Example: An automobile at a stoplight moves forward and reaches a speed of 25 m/s in 15 seconds. What is the average acceleration, assuming travel in a straight line?

Solution: Applying the definition, we get:

$$\overline{a} = \frac{25 \text{ m/s} - 0 \text{ m/s}}{15 \text{ s}} = 1.67 \text{ m/s}^2$$

Note: In this and a number of subsequent examples, we are calculating only the magnitude of various vector quantities. To indicate magnitude, the symbol for that quantity is printed in italic type, not boldface. Hence, we used $\overline{a}$, not $\overline{\mathbf{a}}$, in this example.

2) The *instantaneous acceleration*, **a**, is defined in a manner analogous to the instantaneous speed.

If the direction of the velocity changes, there is an acceleration even if the magnitude of the velocity remains constant. This is a consequence of the definition of acceleration; it is present when there is *any* change in the velocity.

For simplicity we usually drop the words "instantaneous" and "average" when describing motion. It will be evident from context which type is meant.

One-Dimensional Motion

If an object moves in a straight line with a constant acceleration, then there are simple relationships between distance, speed, and acceleration. The simplicity stems from the scalar nature of the variables; direction does not change in one-dimensional motion. Using subscripts i and f to indicate initial and final values, we get the following equations:

1) $v_f = v_i + at$
2) $\overline{v} = \sqrt{\dfrac{v_i + v_f}{2}}$
3) $d = v_i t + \frac{1}{2} at^2$

MOTION

Example: A box is pushed from rest until, after 5 seconds, its speed is 20 m/s. Assume the acceleration is constant. Find:
 a) the average speed over the interval
 b) the magnitude of the acceleration
 c) the distance traveled

Solution: First we list the information given in the problem:

$$v_i = 0$$
$$v_f = 20 \text{ m/s}$$
$$t = 5 \text{ s}$$

For a)
$$\bar{v} = \frac{0 + 20 \text{ m/s}}{2}$$
$$= 10 \text{ m/s}$$

For b)
$$v_f = v_i + at$$
$$20 \text{ m/s} = 0 + a \times 5 \text{ s}$$
$$a = \frac{20 \text{ m/s}}{5 \text{s}} = 4 \text{ m/s}^2$$

For c)
$$d = 0 + \tfrac{1}{2}(4 \text{ m/s}^2)(5 \text{ s})^2$$
$$= 2 \text{ m/s}^2 \times 25 \text{ s}^2 = 50 \text{ m}$$

NEWTON'S LAWS

Newton's laws are laws concerning forces and their relation to motion. For now, we will define a *force* as a push or a pull. Since pushing or pulling sometimes causes an object to move, a force is often thought of as that which causes motion.

Newton's First Law

Newton's First Law states that an object moving at a given velocity maintains that velocity unless acted upon by an unbalanced force. In other words, if the object was originally still, it remains still unless something pushes or pulls it to get it moving. If it was originally moving, it continues moving with its original velocity unless a force alters its motion.

Newton's Second Law

Newton's Second Law relates force, mass, and acceleration as follows:

$$f = ma$$

In this formula, f represents the unbalanced force on an object. When all the forces on an object add up to zero, there is no acceleration. The object is said to be in *equilibrium*. The SI units of force are technically kilogram-meters per second squared, kg-m/s^2. Since this is cumbersome, we define a new unit called a *newton* (N):

$$1 \text{ N} = 1 \text{ kg-m/s}^2$$

Hence, the SI units of force are newtons. In the cgs system, the units are g-cm/s^2, called *dynes*.

Example: If the automobile in the example on page 6 has a mass of 1.5 kg, what is the magnitude of the force needed to accelerate it at the above rate?

Solution: Since we have already determined the acceleration, we simply apply Newton's law:

$$f = (1.5 \text{ kg}) \times (1.67 \text{ m/s}^2)$$
$$= 2.5 \text{ kg-m/s}^2 = 2.5 \text{ N}$$

Newton's Third Law

Newton's Third Law states that if object A exerts a force on object B, then B must exert a force on A equal in magnitude but opposite in direction. This pair of forces is called an *action-reaction pair*.

This law is often misunderstood. If for each force in the world there is an equal and opposite one, why does any motion occur at all? The reason is that action-reaction pairs act on *different* objects. Consider the motion of a car. The tires of a car exert a rearward force on the ground. By Newton's Third Law, the ground exerts an equal force on the car but in the forward direction. The car moves because the force

MOTION

due to the ground (the unbalanced force) pushes it forward. While it is true that the sum of the action plus reaction forces adds up to zero, the sum of the forces *on the car* does not add to zero. Hence, the car moves.

Newton's Law of Gravitation

Any two objects in the universe attract each other by virtue of their mass. Newton's *Law of Gravitation* expresses the magnitude of this force mathematically:

$$f = G \frac{m_1 m_2}{r^2}$$

Here, m_1 and m_2 are the masses and G is a number which has the same value throughout the universe, 6.67×10^{-11} m^3/kg-s^2. If the objects are spherical, which we will assume for simplicity, then r represents the distance between their centers.

Example: What is the magnitude of the force of attraction between two 3 kg watermelons sitting 0.25 m apart on a supermarket shelf?

Solution: Using Newton's Law of Gravitation:

$$f = \frac{(6.67 \times 10^{-11} \text{ m}^3/\text{kg-s}^2) \times 3 \text{ kg} \times 3 \text{ kg}}{(0.25 \text{ m})^2}$$

$$= \frac{6.67 \times 3 \times 3}{.25 \times .25} \times 10^{-11} \times \frac{\text{m}^3 \times \text{kg} \times \text{kg}}{\text{m}^2 \times \text{s}^2 \times \text{kg}} = 960 \times 10^{-11} \frac{\text{kg-m}}{\text{s}^1}$$

$$= 960 \times 10^{-11} \text{ N} = 9.6 \times 10^{-13} \text{ N}$$

This is a tiny force. The gravitational attraction between everyday objects is negligible.

It should be emphasized that the force f is the force exerted by each body on the other. These are action-reaction pairs. In many cases, one object is huge (e.g., the earth) while the other is very small by comparison (e.g., a car). For the earth-car example, we are usually interested only in the motion of the car. Then we need consider only the force of the earth on the car; we can ignore the car's force on the earth.

Simple Applications of Newton's Laws

1) *Free fall* refers to the motion of an object acted upon only by the gravitational attraction of the earth; air resistance is excluded. In ana-

lyzing free fall, we consider only the region near the surface of the earth. In that case, the distance between the object and the center of the earth is approximately the radius of the earth.

All objects falling freely accelerate at the same rate. This rate is called the *acceleration of gravity*, g, and has a magnitude of 9.8 m/s² near the earth's surface. We can see this from Newton's laws. For a body, b, let us find the acceleration, a_b. According to the Second Law, the force on b is:

$$f_b = m_b a_b$$

According to the Law of Gravitation, the force on b is:

$$f_b = G m_e m_b / R^2$$

Now just set the two expressions equal:

$$m_b a_b = G m_e m_b / R^2$$

Solving for the acceleration, we get:

$$a_b = G m_e / R^2$$

Note that the acceleration does not depend on the mass of the body, just on the radius and mass of the earth. If we plug in those quantities, we get the value for g.

Example: What is the magnitude of g for the moon? Use 7.4×10^{22} kg for the mass and 1.74×10^6 m for the radius.

Solution: Using the formula just derived:

$$g_m = \frac{(6.67 \times 10^{-11} \text{ m}^3/\text{kg-s}^2) \times (7.4 \times 10^{22} \text{ kg})}{(1.74 \times 10^6 \text{ m})^2}$$

$$= \frac{(6.67 \times 7.4)}{1.74^2} \times \frac{(10^{-11} \times 10^{22})}{10^{12}} \times \frac{\text{m}^3 \times \text{kg}}{\text{m}^2 \times \text{s}^2 \times \text{kg}}$$

$$= 1.63 \text{ m/s}^2$$

Example: How long would it take for an apple to fall from the top of a tree 10 m tall?

Solution: Use the linear motion formula $d = v_i t + \frac{1}{2} a t^2$.

$$d = 10 \text{ m}$$
$$v_i = 0$$
$$a = 9.8 \text{ m/s}^2$$

MOTION

$$\text{Then: } 10 \text{ m} = \frac{1}{2}(9.8 \text{ m/s}^2)t^2$$
$$(20/9.8)\text{s}^2 = t^2$$
$$t = 1.43 \text{ s}$$

2) The *weight* of an object is the force exerted on it by gravity. By Newton's Second Law, we can express the magnitude of the weight as:

$$w = mg$$

From our discussion of acceleration of gravity, we know that the magnitude of g varies from planet to planet. Thus, our weight is different on different planets even though our mass is the same. On the moon we weigh one-sixth as much as here on earth. This is why astronauts can jump to such great heights on the moon. Their leg muscles are strong enough to cope with the much stronger earth gravity; hence, they find the pull of the moon very easy to overcome.

Example: What is the magnitude of the weight of the 3 kg watermelon of the previous example?

Solution: Weight = $(3 \text{ kg}) \times (9.8 \text{ m/s}^2)$
 = 29.4 N

Note how much larger this is than the gravitational attraction between the watermelons themselves.

3) *Friction* is a phenomenon which refers to the action-reaction forces involved when two objects in contact move relative to each other. Consider a box which we start to push along a horizontal floor. *Static friction* is the horizontal force exerted by the floor and the box on each other while they are still. This force can vary. When we start to push the box with a weak force, the box doesn't move. This is because the weak pushing force is opposed by the force of static friction. As we slowly increase our pushing force, the static friction force increases. However, there is a maximum beyond which the static friction force cannot increase. Once the pushing force exceeds this value, the box moves. *Kinetic friction* is the force exerted between the floor and the box while the box moves. To keep the box moving, we have to exert enough force to overcome the force of kinetic friction. However, this is less than that required to get the box moving initially. Static friction is always greater than kinetic friction for a given pair of surfaces.

The frictional force depends on the nature of the surfaces in contact and the weight of the box. For instance, if the box and the floor were made out of rubber, we would have a hard time pushing the box. Also, if the box were simply heavier, it would be harder to push.

TWO-DIMENSIONAL MOTION

There are important examples of motion that can be regarded as taking place in a particular plane. Although this is more complicated than one-dimensional motion, no new motion laws need be introduced. All we have to do is treat the vertical and horizontal components separately.

Example: How far and high does a shell go if it is fired at a 45° angle from a cannon? Assume the initial speed, both vertically and horizontally, is 50 m/s.

Solution: We see from figure 1-2 that the shell rises to a height H and falls back to the ground while traversing a horizontal distance d. These two quantities can be calculated as follows:

1) Calculate the time required to reach the height H by using the formula $v_{f_y} = v_{i_y} + a_y t$ (y indicates vertical direction):

$$v_{f_y} = 0$$
$$a_y = -9.8 \text{ m/s}^2 \text{ (Note minus sign!)}$$
$$v_{i_y} = 50 \text{ m/s}$$

We get:

$$0 = 50 \text{ m/s} - 9.8 \text{ m/s}^2 \times t$$
$$\text{or}$$
$$50 \text{ m/s} = 9.8 \text{ m/s}^2 \times t$$

Dividing by 9.8 m/s² we get:

$$t = \frac{50 \text{ m/s}}{9.8 \text{ m/s}^2} = 5.1 \text{ s}$$

2) The same time is required to fall back to earth. Hence the total time in the air is 10.2 seconds.

3) Since the shell spent 10.2 seconds in motion, we can figure out how far it went by using $d = v_{i_x} t + 1/2 \, a_x t^2$ (x indicates horizontal direction)

$$v_{i_x} = 50 \text{ m/s}$$
$$t = 10.2 \text{ s}$$
$$a_x = 0$$

MOTION

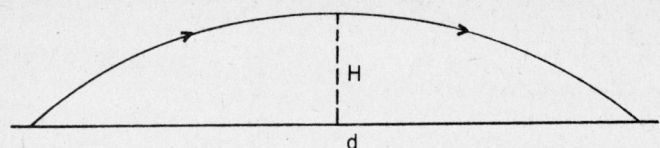

Fig. 1-2 Path of an artillery shell

Remember, there is no force on the shell in the horizontal direction, so a_x is zero. Thus, $d = v_{i_x} t$. Solving for d, we get:

$$d = 510 \text{ m}$$

It turns out that the 45° angle gives the largest value of d for a given initial speed.

UNIFORM CIRCULAR MOTION

If an object moves in a circle, the radius sweeps out a certain angle, θ. In this section we will investigate motion from the point of view of angular changes.

Radian Measure

One *radian* is the angle subtended by an arc of a circle whose length is the same as the radius (figure 1-3). In general, the angle in radians is the arc length (d) divided by the radius (r):

$$\theta = d/r$$

For instance, if the arc were half the radius, the angle would be half a radian. When using the above formula, the arc length and the radius must be in the same units. Hence, the radian is a unitless quantity. In analyzing circular motion, the angle must always be measured in radians, not degrees.

Example: How many radians are there in one revolution?

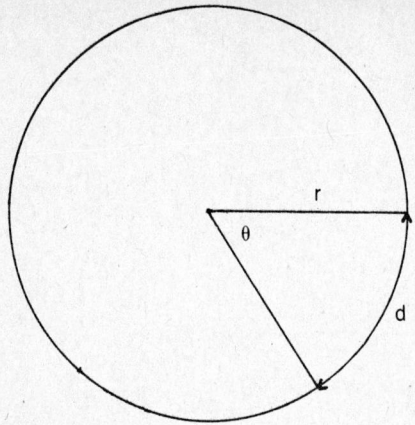

Fig. 1-3 Radian measure

Solution: In one revolution, the arc length is the circumference of a circle. Hence:

$$\theta = 2\pi r/r = 2\pi = 6.28 \text{ radians}$$

Angular Speed

1) The *average angular speed* in an interval, ω, is the number of radians swept out divided by the time elapsed:

$$\omega = \frac{\Delta\theta}{\Delta t}$$

Example: What is the angular speed of a long-playing record?

Solution: The record turns 33 1/3 revolutions per minute. First we calculate the number of radians:

$$33\ 1/3 \text{ rev} \times 6.28 \text{ rad/rev} = 209.3 \text{ radians}$$

Then we divide by 60 seconds to get the angular speed:

$$\omega = 209.3 \text{ radians}/60 \text{ s} = 3.49 \text{ rad/sec}$$

Since "radians" is unitless, we can leave it off and write the units of ω as $1/s$ or s^{-1}.

MOTION

2) The *instantaneous angular speed* can be thought of as the angular speed at a given moment.

Angular Velocity

Angular velocity is a vector, ω, whose magnitude is the angular speed and whose direction is perpendicular to the plane of motion. The direction is such that an object moving counterclockwise on this page has an angular velocity pointing out of the page. In this book, we will consider only cases where the angular velocity is constant. This is why we have not made a distinction between average and instantaneous angular velocity.

Linear Quantities in Circular Motion

1) The actual distance an object travels when moving in a circle is just the angle swept out multiplied by the radius:

$$d = r \times \theta$$

Note that this formula comes from the definition of the radian.

Example: How far does a point on the outer edge of a long-playing record travel in one minute? Assume a radius of 0.15 m.

Solution: In one minute the record travels 209.3 radians. Using the above formula, we get:

$$d = 0.15 \text{ m} \times 209.3 \text{ radians}$$
$$= 31.4 \text{ m}$$

2) The linear speed is just the angular speed multiplied by the radius:

$$v = r \times \omega$$

Example: How fast does the edge of the long-playing record move?

Solution: Since we have already figured out ω, and r = .15 m, we can apply the above formula:

$$v = 0.15 \text{ m} \times 3.49 \text{ s}^{-1}$$
$$= 0.52 \text{ m/s}$$

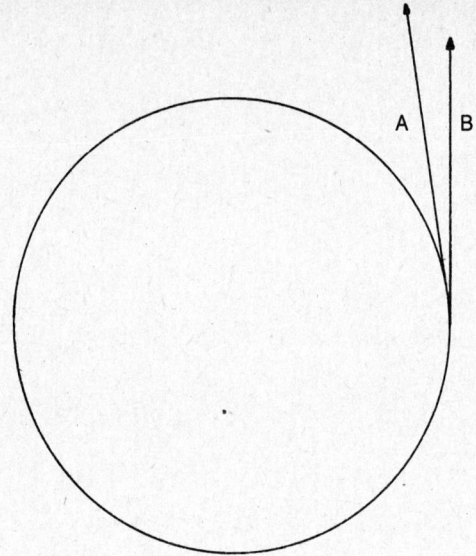

Fig. 1–4 Velocity vectors for an object in circular motion

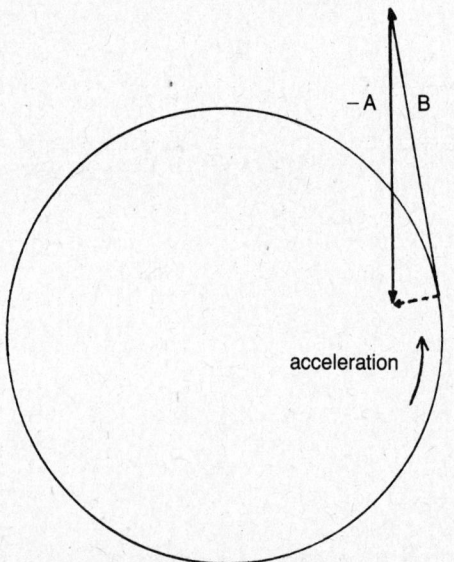

Fig. 1–5 Acceleration vector for circular motion. The acceleration points in the direction of the difference of the two velocity vectors.

MOTION

3) The *linear velocity vector* for an object moving in a circle always points tangent to the circle. This is illustrated in figure 1-4 for two instants of time.

4) The linear acceleration vector, called *centripetal acceleration,* points toward the center, as shown in figure 1-5. This is because the acceleration points in the direction of the vector difference between final and initial velocities. Figure 1-5 shows the result of the vector subtraction for the two vectors of figure 1-4. The result clearly points toward the center. The magnitude of the centripetal acceleration is v^2/r.

Example: A small bug rests at the edge of our record. What is its acceleration?

Solution: It experiences a centripetal acceleration whose magnitude is:

$$a = (0.52 \text{ m/s})^2 / 0.15 \text{ m}$$
$$= 1.8 \text{ m/s}^2$$

Moment of Inertia

The *moment of inertia, I,* is a measure of an object's resistance to changes in its angular speed. An object with a large moment of inertia resists changes in angular speed more than one with a small moment of inertia. This is analogous to mass for the linear case: large masses resist changes in their linear speed more than small ones.

The moment of inertia depends on the mass of the object and, more importantly, on how the mass is distributed relative to the axis of rotation. If most of the mass is concentrated near the axis, I is relatively small. If most of the mass is concentrated far from the axis, I is relatively large.

Torque

The *torque* on an object is a quantity that tends to make the object rotate about an axis. It is the product of a force times the distance from the line of action of the force to the axis of rotation. The seesaw, figure 1-6, illustrates these ideas. Box A exerts a force f on one side of the seesaw. This results in a torque, $f \times d$, which tends to make the seesaw rotate clockwise. Similarly, box B exerts a torque $f' \times d'$ which

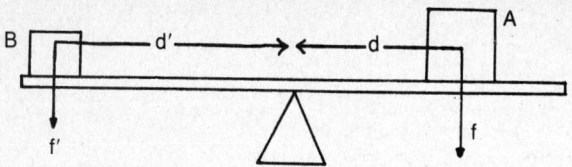

Fig. 1–6 Torques on a seesaw. For balance we must have $fd = f'd'$.

tends to make the seesaw rotate counterclockwise. The seesaw will balance if the two torques are equal in magnitude but opposite in sense. Note that the forces do not have to be equal.

Example: Two children of weights 270 N and 360 N get on a seesaw. The heavier one sits 3 m from one end. Where must the other child sit in order to balance the seesaw?

Solution: This is the same situation as figure 1-6. The torque exerted by the heavier child is:

$$\text{torque} = 360 \text{ N} \times 3 \text{ m} = 1080 \text{ N-m}$$

The lighter child must exert the same torque at some distance d:

Solving for d we get:
$$1080 \text{ N-m} = 270 \text{ N} \times d$$
$$1080 \text{ N-m}/270 \text{ N} = 4 \text{ m}$$

The child must sit 4 m from the end on the other side.

Centripetal and Centrifugal Force

Centripetal force on an object in circular motion is a force directed toward the center whose magnitude is the centripetal acceleration times the mass.

Centrifugal force is an apparent outward force on an object in circular motion. This requires an involved explanation.

Imagine a turntable that has a raised outer edge and rotates at a constant velocity. On this turntable, resting up against the edge, is a ball. An observer located off the spinning turntable (in a fixed frame of ref-

MOTION

erence) sees the ball in uniform circular motion. There is one force on the ball: the inward (centripetal) force provided by the raised edge.

An observer on the turntable is in a rotating frame of reference and sees things differently. He maintains that the ball is in equilibrium, since it is not moving with respect to him. If he moves the ball toward the center and lets go, the ball moves outward toward the edge. He concludes that there is an outward, or *centrifugal,* force on the ball. He explains the equilibrium at the edge as a balance between the centrifugal and centripetal forces.

Why the disparity in observations? Consider figure 1-7. After the ball is let go at point P, there is no force on it. Hence, it travels in a straight line *(PQ)* until it hits the edge. The observer, on the other hand, always moves in a circle (arc *PR*). The pairs of dots on the radii indicate the relative position of the ball and observer at four different times after the ball is released. It should be clear that if the observer looks out along a radius, he always sees the ball along that radius. The ball appears only to get farther away until it hits the edge. (For simplicity, we will assume that upon hitting the edge, it resumes its circular motion.) The observer thus concludes that there is a radially outward force on the ball, even though there is none.

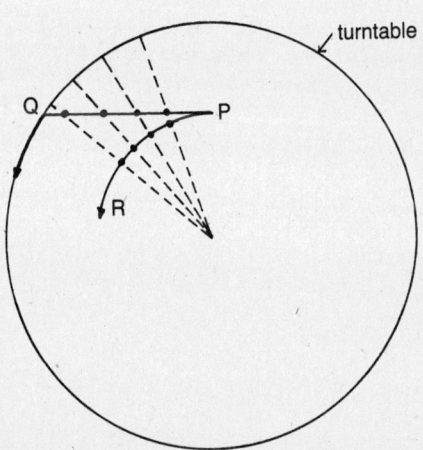

Fig. 1–7 Rotating coordinate system

We see that in a rotating frame of reference an object can behave as if there is a force on it, even if there really isn't. The remarkable thing is that if we want to use Newton's laws in a rotating frame, we must make believe that the imagined force actually exists. The observer on the turntable, for instance, can account for the apparent equilibrium of the ball at the edge *only* by concluding that there is a centrifugal force (imagined) to balance the real centripetal force. If only the centripetal force is considered, then it would appear that the ball is in equilibrium under the influence of an unbalanced force. This violates Newton's laws. Note that the observer in the fixed frame doesn't claim that the ball is in equilibrium. The unbalanced force doesn't present any problems for him.

It is important to realize that the moving observer's description of the motion is just as valid as that of the fixed observer. Although we usually analyze motion from a fixed frame of reference, it is sometimes advantageous to use a rotating frame. For instance, if you are in a car making a sharp left turn, it is natural to use the moving car as the frame of reference. Then you would describe yourself as being pushed to the right by a force. Later in this text you will see that a rotating frame of reference is a convenient one to use in describing two important phenomena, tides and winds.

2
Momentum, Work, and Energy

In this chapter, we expand our study of motion by introducing concepts such as momentum and energy. These are necessary to give a more complete description of motion than we have seen up to now.

MOMENTUM

Linear Momentum

The linear momentum, **p**, is essentially a measure of a body's tendency to stay in motion. Mathematically, we express momentum as:

$$\mathbf{p} = m\mathbf{v}$$

Hence, the tendency to stay in motion is great if either the mass or the velocity is large.

The *Law of Conservation of Momentum* states that the momentum of a system is constant if no external forces act on it. The *system* is the collection of objects whose motion we are considering. *External forces* are those which are not exerted by one part of the system on the other. The momentum of the system is just the vector sum of the momenta of all parts of the system. Conservation of momentum is a powerful tool which we can use to figure out the motion of a system. Consider the following situations.

1) Imagine a rocket ship at rest out in space. Assume it is far from any planet so there is no gravitational force on it (i.e., no external force). The rocket and its fuel comprise the system; since they are at rest, the total momentum of the system is zero. When the engine fires,

the fuel develops a momentum in the rearward direction. In order that the total momentum of the system remain zero, the rocket must develop a momentum in the forward direction equal in magnitude to the rearward momentum of the fuel. Thus the rocket moves forward.

2) A gun which is fired recoils backward. Before firing, the system (gun + bullet) has a momentum of zero. By conservation of momentum, its momentum after firing must still be zero. After the trigger is pulled, the bullet develops a forward momentum. Hence, the gun must develop an equal rearward momentum.

Example: A gun can fire a 0.05 kg bullet at a speed of 1000 m/s. If the gun has a mass of 10 kg, at what speed does it recoil?

Solution: The magnitude of the forward momentum of the bullet is

$$(1000 \text{ m/s}) \times (0.05 \text{ kg}) = 50 \text{ kg-m/s}$$

The magnitude of the rearward momentum of the gun must be the same:

$$50 \text{ kg-m/s} = 10 \text{ kg} \times u$$

We can solve for u: $u = 5$ m/s

Angular Momentum

The *angular momentum*, **L**, of a rotating body is the product of moment of inertia and angular velocity:

$$\mathbf{L} = I\omega \quad (\text{compare: } \mathbf{p} = m\mathbf{v})$$

It is a measure of a body's tendency to remain spinning. Angular momentum follows a conservation law analogous to that for linear momentum: it remains constant unless an external torque acts on the body. For instance, consider a figure skater spinning freely with her arms at her side. She can slow her rotational speed by holding both arms straight out. This increases the moment of inertia because more mass is distributed farther away from the axis of rotation. The higher moment of inertia, I, must be balanced by a lower rotational velocity, ω, in order that the product, $I\omega$, remain constant.

MOMENTUM, WORK, AND ENERGY

WORK AND POWER

Work

In physics, *work* is done when a force acts over a distance. It is calculated by a simple formula:

$$w = f \times d$$

The SI units are newtons times meters, N × m. One newton-meter is called a *joule*. In the cgs system, the units are dynes × cm, or ergs. This formula can be applied only under two conditions:

1) the force must be constant over the distance involved.
2) the force must be in the direction of motion.

Let us investigate the work done for two cases when a child pulls a sled along the ground. We will assume the force f exerted by the child is constant. In figure 2-1, the force is in the direction of motion. If the sled moves a distance d, then the work done is $f \times d$. In figure 2-2, the force is not in the direction of motion; $f \times d$ gives the wrong answer.

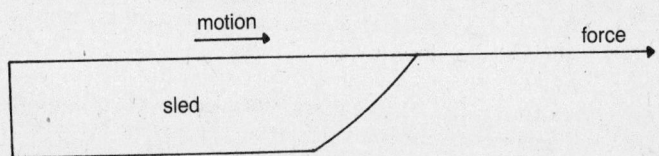

Fig. 2–1 Sled pulled by a force in the direction of motion

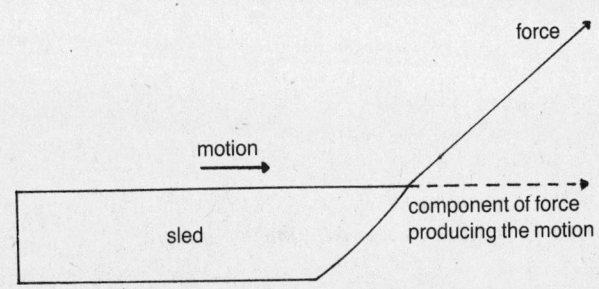

Fig. 2–2 Sled pulled by force *not* in the direction of motion

However, f can be resolved into components f_x and f_y (see Appendix). Since only f_x is in the direction of motion, the work done is $f \times d$.

Example: How much work is done when a 50 kg box is lifted from a shelf 1 m high to a shelf 3.5 m high?

Solution: The force needed to lift the box is its weight.

$$\text{Force} = \text{weight} = 50 \text{ kg} \times 9.8 \text{ m/s}^2$$
$$f = 490 \text{ N}$$

Since this force is in the direction of motion, we can apply the formula for work:

$$\text{Work} = 490 \text{ N} \times 2.5 \text{ m} = 1225 \text{ joules}$$

Power

Power is the rate at which work is done:

$$P = \text{work}/\text{time} \text{ (units: joules/second, called watts)}$$

Power is important because it is often useful to know how much work a machine can do in a certain time. One *horsepower* is just a convenient unit for denoting the maximum amount of work per unit time that a motor or engine can perform. It is equal to 746 watts.

Example: How many watts of power are needed to do the work of the previous example in 7 seconds?

Solution: According to the formula,

$$P = 1225 \text{ joules}/7 \text{ s}$$
$$= 175 \text{ watts}$$

ENERGY

Energy is the capacity to do work. There are many forms of energy. In this section, we will consider two: kinetic and potential.

MOMENTUM, WORK, AND ENERGY

Kinetic Energy

Kinetic energy, *KE,* is the energy a body possesses because it is moving. It is expressed mathematically as:

$$KE = 1/2\ mv^2 \text{ (units: joules)}$$

Example: What is the kinetic energy of a 1000 kg car moving at 10 m/s?

Solution: Kinetic energy = $1/2\ (1000\text{ kg}) \times (10\text{ m/s})^2$
  = 50,000 kg-m^2/s^2
  = 50,000 joules

In order to increase the kinetic energy of a body, an amount of work equal to the increase must be done on the body.

Example: How much work must be done in order to increase the speed in the previous example to 20 m/s?

Solution: Kinetic energy at 20 m/s = $1/2\ (1000\text{ kg}) \times (20\text{ m/s})^2$
  = 200,000 joules

We must therefore do 150,000 joules of work in order to increase the car's speed to 20 m/s. This illustrates why it normally is most economical to drive at a constant speed. At a steady speed, the kinetic energy remains constant. Hence (if there are no hills), the engine uses gas only to overcome friction, not to increase the kinetic energy.

Potential Energy

Potential energy is the energy that a body has because of its position. For instance, a body high off the ground has the potential to do work. It can fall and the resulting motion can be harnessed to do work. The potential energy a body has because of height is called *gravitational potential energy*. We can illustrate some important ideas about potential energy by discussing gravitational potential energy.

A body at a particular height does not have a unique amount of potential energy. Figure 2-3 illustrates why. A block at position *A* can fall to either position *B* or *C.* Since the block gains speed as it falls, it is

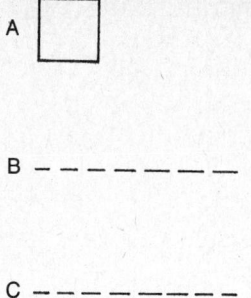

Fig. 2-3 Potential energy of block depends on the choice of B or C as the zero level.

clear that in falling to C it gains more speed than in falling to B. Hence, the amount of work it can do depends not only on the initial height but also on how far it falls. Despite this complication, it is convenient to associate a definite value of potential energy with each height. This is done by choosing a reference level in a problem and assigning that level the value zero PE. Then, for a height h above the reference level, the potential energy is defined as:

$$PE = mgh \text{ (units: joules)}$$

Although PE defined in this way seems arbitrary, we will see in the next section that it doesn't really matter. This is because the physically meaningful quantity is really the *difference* in PE between two points, not the actual values.

Example: Assume that the floor of a room is the zero of PE. What is the PE of a 50 kg box located on a shelf 1.0 m high?

Solution: From the formula, the PE is:

$$PE = (50 \text{ kg}) (9.8 \text{ m/s}) (1.0 \text{ m})$$
$$= 490 \text{ joules}$$

As with kinetic energy, it takes work in order to increase the potential energy.

Example: What is the change in PE when a 50 kg box is lifted from a shelf 1.0 m high to a shelf 3.5 m high?

MOMENTUM, WORK, AND ENERGY

Solution: PE at 1.0 m = 490 joules.
PE at 3.5 m = (50 kg) (9.8 m/s^2) (3.5 m)
            = 1715 joules
PE difference = 1715 joules − 490 joules = 1225 joules

This is exactly the work done in lifting it to the higher shelf.

TRANSFORMATION OF ENERGY

By *transformation of energy,* we mean the change of energy from one form to another. When energy changes take place, they obey the law of *Conservation of Energy:* the total energy of a system remains constant. We can see what this means by examining a simple device, the pendulum.

Figure 2-4 shows a pendulum that has been pulled up a height h from its lowest position. We have assigned the lowest position the value $PE = 0$. In the position shown, the pendulum has a potential energy of mgh; its kinetic energy is zero because initially it is not moving. The total energy in this position is therefore mgh. Because of conservation of

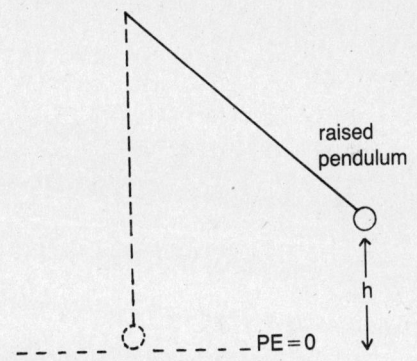

Fig. 2-4 Pendulum raised to height h above zero level

energy, its total energy will be the same in *all* positions. Hence, for this problem, we can write:

$$\text{total energy} = mgh$$

As it moves toward its lowest position, its *PE* decreases. Its *KE* must therefore increase so that the total remains constant. At the lowest position, the *PE* is zero and all of the pendulum's energy is kinetic, making its speed a maximum there. Since the total energy for this pendulum is always *mgh,* we can write the following energy equation for the pendulum at the lowest point:

$$mgh = 1/2\, mv^2 + 0$$
$$(\text{total}) = (\text{kinetic}) + (\text{potential})$$

This equation can be solved for v:

$$v = \sqrt{2gh}$$

This is the maximum speed of the pendulum; quite reasonably, it depends on how high it is pulled up.

What if we had defined the zero of potential as in figure 2-5? The final result would still be $v = \sqrt{2gh}$. Here's why. The pendulum is pulled aside a height $(h + H)$. Its initial potential energy is therefore $mg(h + H)$. Its initial kinetic energy, as before, is zero. Hence, for this problem we can write:

$$\text{total energy} = mg(H + h)$$

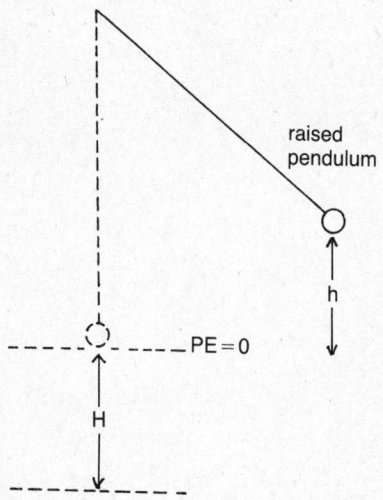

Fig. 2–5 Pendulum raised to height $h + H$ above zero level

MOMENTUM, WORK, AND ENERGY

The *PE* at the lowest point is not zero but *mgH*. The *KE* is still $1/2\,mv^2$. The equation for the lowest point is now:

$$mg(h + H) = 1/2\,mv^2 + mgH$$
$$(\text{total}) = (\text{kinetic}) + (\text{potential})$$

Since the left side can be written as *mgh + mgH*, the term *mgH* drops out from both sides. The equation reduces to the same equation we had when we assumed the zero was at the lowest part of the swing. Thus it doesn't matter where the zero of potential is chosen. What counts is the *PE* difference between the lowest and highest levels. That difference is independent of reference level.

What happens to the pendulum after it reaches the bottom of its swing? Its momentum carries it past the bottom and back up again. It gains *PE* and loses *KE*. Eventually, it loses all of its *KE* and reaches a point where its energy is all potential. Since the total energy is constant, this value of *PE* must be the same as the original. Hence, the height it reaches on the upswing must be the same as the original height.

In this problem, we neglected friction. Without friction, the energy the system has by virtue of its initial height is transformed back and forth *indefinitely* from kinetic to potential. The effect of friction is to constantly dissipate energy. On any given upswing, the pendulum does not have enough energy to reach the height it attained on the previous upswing. Each swing is thus shorter and shorter. Eventually, no energy is left, and the pendulum comes to rest.

3
Bulk Matter

All matter is made of microscopic building blocks called atoms and molecules. We will study atoms and molecules in detail later. For now it is sufficient to know that these particles exert forces on each other which enable matter to exist in bulk. In this chapter, we will consider the general properties of bulk matter and investigate one aspect, fluids, in detail.

STATES OF MATTER

There are three states of matter: solid, liquid, and gas. Sometimes gases and liquids are lumped together and called *fluids*. The distinguishing characteristics of each state are determined by the strength of the attracting force between neighboring molecules.

Solids

Solids are substances that have a definite shape and volume. However, their shape can be deformed when an external force is applied. Some solids, such as rubber, are *elastic:* they return to their original shape after the deformation. Others, such as crystals, are *brittle:* they shatter when deformed. A few, like gold, are *malleable:* they can be hammered out into extremely thin sheets. In a solid, the attractive force between neighboring molecules is the strongest of the three states.

BULK MATTER

Liquids

Liquids are substances which have a definite volume but no definite shape. They assume the shape of their container. Liquids can flow, but some flow more easily than others. The resistance to flowing is called *viscosity*. In a liquid, the attractive force between neighboring molecules is weaker than for the solid state. This allows liquid molecules to slip past one another.

Gases

Gases have neither definite shape nor definite volume. They can expand to fill any volume or be compressed into very small volumes. The attractive force between molecules of a gas is extremely weak.

Change of State

Changes between the three states have names which depend on which two states are involved.

1) *Melting* occurs when a solid becomes a liquid.
2) *Evaporation* occurs when a liquid becomes a gas.
3) *Condensation* is the opposite of evaporation: a gas becomes a liquid.
4) *Freezing* occurs when a liquid becomes a solid.
5) *Sublimation* occurs when a solid becomes a gas.

DENSITY

The *density* of a substance is the mass divided by the volume:

$$d = m/v \text{ (units: kg/m}^3\text{)}$$

For a given solid or liquid, the density is constant. For gases, the density varies because gases have no definite volume. Knowing the density and the volume allows us to compute the mass; knowing the density and the mass allows us to compute the volume.

Example: What is the mass of 3.5 m³ of aluminum, assuming the density of aluminum is 2.7×10^3 kg/m³?

Solution: Using $d = m/v$ we have:

$$2.7 \times 10^3 \text{ kg/m}^3 = m/3.5 \text{ m}^3$$

Multiplying, we get:

$$(3.5 \text{ m}^3) \times (2.7 \times 10^3 \text{ kg/m}^3) = m$$

$$9.45 \times 10^3 \text{ kg} = m$$

Example: What is the volume of 5.4 kg of aluminum?

Solution: Again using $d = m/v$:

$$2.7 \times 10^3 \text{ kg/m}^3 = 5.4 \text{ kg}/m$$

$$m = 5.4 \text{kg}/(2.7 \times 10^3 \text{ kg/m}^3)$$

Solving, we get:

$$m = 2 \times 10^{-3} \text{ m}^3$$

PRESSURE

The *average pressure*, $\overline{P}$, over an area, A, is the total force, f, acting perpendicular to the area divided by the area:

$$\overline{P} = f/A \text{ (SI units: N/m}^2\text{, called } pascals, \text{ pa)}$$

We must use the term average because the pressure can vary over the area. However, if we consider an area which is microscopically small, then we can speak of the pressure at a point. Pressure in a fluid is a very important quantity because fluids flow in response to changes in pressure. Solids do not.

For fluids at rest, two important principles hold:

1) The pressure at a given point is the same in all directions. If it weren't, the fluid would flow and no longer be at rest.

BULK MATTER

2) Pressure varies only with the depth below the surface and not with other factors such as the shape of the vessel. This is because pressure over an area is due simply to the weight of the fluid above that area. The pressure at a depth h due to the fluid is:

$$P = hdg \ (d = \text{density})$$

Atmospheric Pressure

We live at the bottom of a fluid called the *atmosphere*. At the surface of the earth, the pressure due to all that air above us is 10^5 pa. In more familiar terms, every square inch has a force of about 15 pounds on it because of the atmosphere. The roof of a house, for instance, has a tremendous force pushing down on it due to the atmosphere. However, because of principle (1), an equal force pushes up, canceling out the downward force.

Barometer

A *barometer* is a device used to measure atmospheric pressure. As seen in figure 3-1, a long evacuated tube, closed at one end, is inverted into a pool of mercury. The force exerted by the atmosphere pushes the mercury into the tube. The mercury rises until the weight of the mer-

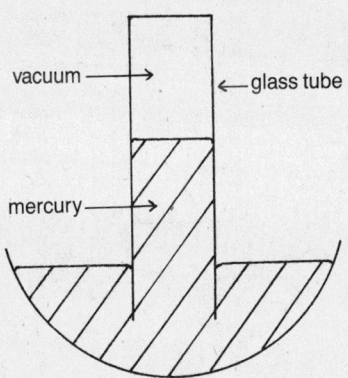

Fig. 3–1 Mercury barometer

cury in the tube just balances the force pushing it up. The height of the mercury is thus a good indicator of the atmospheric pressure.

FLUID BEHAVIOR

Buoyant Force

Buoyant force is the upward force exerted by a fluid on any object immersed in it. The reason for its existence is indicated in figure 3-2. The pressure at the top of the box, due to the fluid, is dgh. At the bottom, the pressure is $dg(h + a)$. The downward force at the top is clearly less than the upward force at the bottom, by an amount dga. The result is a net upward force. Perhaps you have noticed that it is easier to lift someone *in* a swimming pool than *out* of it. This is because part of the necessary upward force is provided by the buoyant force.

Archimedes' Principle

Archimedes' Principle states that the buoyant force on an object is equal to the weight of the displaced fluid. We can see why by considering the box of figure 3-2. The buoyant force depends only on the

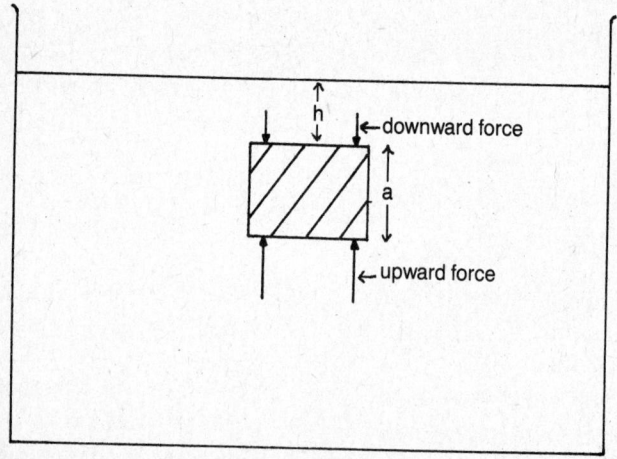

Fig. 3–2 Block in equilibrium in a fluid

BULK MATTER

difference between the pressures at the top and bottom of the box. It does *not* depend on the nature of the box itself. We can make use of this fact by imagining that the box is made out of the fluid itself. Clearly *that* "box" neither sinks nor floats to the top. If it did, any arbitrary section of the fluid which we call our "box" would spontaneously rise or fall, which is ridiculous. Since the "box" remains still, the upward (buoyant) force must be the same as the downward force (weight). Thus the buoyant force is simply the weight of the box of fluid. Even if the box is made out of something else, the buoyant force on it is still the weight of an equal volume of fluid.

Archimedes' Principle helps us to determine whether an object floats or sinks. If the density of the object is greater than the density of the fluid, the displaced fluid weighs less than the object. The buoyant force is smaller than the weight, so the object sinks. If the density of the object is less than the density of the fluid, the displaced fluid weighs more than the object. Thus, the object rises. A floating object with a density significantly less than that of the fluid floats with most of its mass out of the fluid. This is because only a small volume of fluid need be displaced to equal the weight of the object. If a floating object has a density close to that of the fluid, it must have most of its mass below the surface. This is because a large amount of fluid must be displaced in order to equal the weight of the object. This is the case with icebergs. Their density is 90 percent of the density of water. Hence, 90 percent of their mass lies below the surface.

Examples of buoyant force abound. A helium-filled balloon rises because it weighs less than the air it displaces. A ship is made out of metal which normally sinks. However, its shape is such that the water it displaces is enormous. The displaced water weighs more than the ship, so the ship floats.

Pascal's Principle

Pascal's Principle states that if additional pressure is applied to any point in a fluid, the same amount of additional pressure appears at all points in the fluid. Consider two points in a fluid at the same depth. We know that the pressure at those two points must be the same. If we suddenly increase the pressure at one point, that increase must show up at the other point, since both must remain at the same pressure. For two points at different depths, there must be a difference of pressure for the

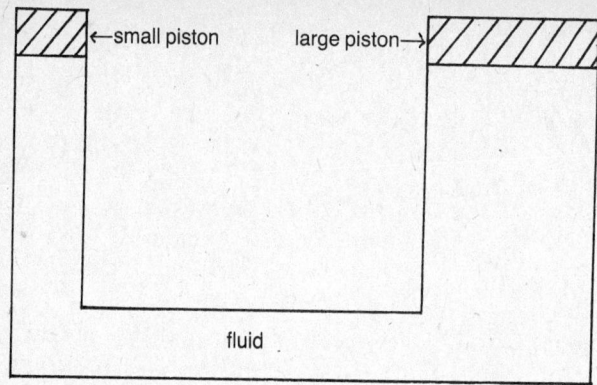

Fig. 3-3 Hydraulic lift

fluid to remain at rest. If we add some pressure at one point, that additional pressure must appear at the other point so that the original difference in pressure can be maintained.

1) A *hydraulic lift* is a good example of Pascal's Principle. Recall the definition of pressure: $P = f/A$. In figure 3-3, we can create a moderately large pressure by exerting an average force over a small area such as that of the left piston. By Pascal's Principle, this pressure must appear at the large piston on the right. This gives rise to an upward force of:

$$f = PA$$

Clearly, if we make the area of the right piston large, the force at the right will be large. Thus, Pascal's Principle helps us to exert a large force without much effort.

Example: A 90 kg person sits on the small piston of figure 3-3. The piston has an area of .006 m². If the large piston has an area of .12 m², how heavy a car can the person lift?

Solution: A 90 kg person has a weight of 882 N. Using the definition of pressure, we can calculate the pressure on the small piston:

$$P = \frac{882}{.006} \text{ N/m}^2 = 147{,}000 \text{ N/m}^2$$

BULK MATTER

This is also the pressure on the large piston. Hence, the force on the large piston is:

$$f = (147{,}000 \text{ N/m}^2)(.12 \text{ m}^2) = 17{,}640 \text{ N}$$

This is a car whose mass is 1800 kg.

2) *Automobile brakes* work by the same principle. From figure 3-4, we see that there is a small *master cylinder* whose fluid chamber is attached to four larger *wheel cylinders*. When the master cylinder is pushed by the piston, the added pressure is transmitted to the four wheel cylinders. Since the total area of the wheel cylinders is several times larger than the area of the master cylinder, the force supplied by them is several times as large as the force with which the brake pedal is pushed.

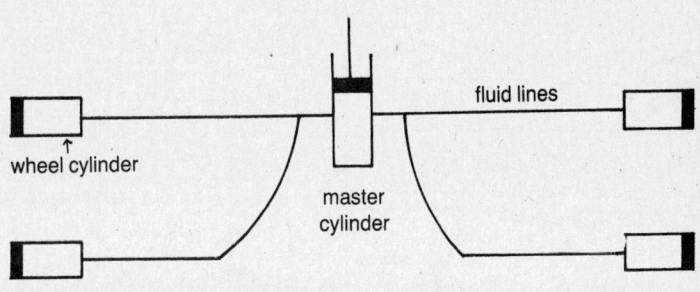

Fig. 3-4 Hydraulic brake system

BERNOULLI'S PRINCIPLE

Bernoulli's Principle is a complex equation involving pressures and energies. Very loosely put, it states that when the velocity of a fluid increases, the pressure decreases.

Figure 3-5 shows the cross section of an airplane wing. Air flowing over the top of the wing has to travel farther, in the same time, than

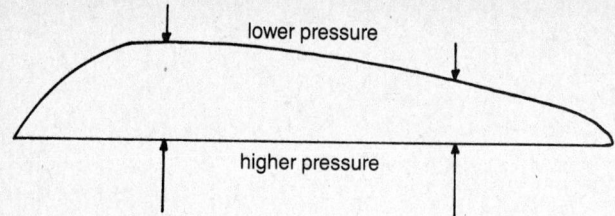

Fig. 3-5 Airplane wing

does air traveling under the bottom. The air at the top thus goes faster, making the pressure at the top lower. The resulting pressure difference creates an upward force which lifts the plane. The same sort of lift is evident to umbrella carriers on a windy day because an umbrella is similar in cross section to a wing.

4
Kinetic Theory

Kinetic theory rests on two main assumptions. The first is that all matter is made up of particles called atoms and molecules. The second is that these particles are in constant motion. The most direct evidence for these ideas comes from so-called *Brownian movement*. This is the name given to the random motion of microscopic particles. For example, one can see (with the aid of a microscope) tiny particles darting about on the surface of a beaker of water. Such particles are buffeted by rapidly moving molecules.

INTERNAL ENERGY

The *internal energy* of a substance is the sum of all the kinetic and potential energies of the atoms and molecules. The sources of these energies are as follows:

(1) Potential energy due to the relative positions of the atoms within a molecule.
(2) Potential energy due to the relative positions of the molecules themselves.
(3) Kinetic energy due to the motion of the atoms within a molecule.
(4) Kinetic energy due to the motion of the molecule as a whole. We refer to this as *translational kinetic energy*.

TEMPERATURE AND HEAT

Temperature is a measure of the *average* translational kinetic energy of the molecules of a substance. Two objects at the same temperature have the same average molecular translational kinetic energy, regardless of the value of other internal energy components. It is an observed fact that two objects at different temperatures placed in contact with each other will eventually reach the same intermediate temperature. This situation is called *thermal equilibrium.*

Heat is the energy that flows between two substances because of a difference in their temperatures. The normal flow of heat is from the hot body to the cooler one. Heat is not something that a body possesses. A body has only internal energy. It is the internal energy which is increased or decreased when heat flows in or out. Heat is associated with a process, not a body.

Thermometers

A *thermometer* is a device having any physical property that changes measurably with temperature. That property is assigned values on an arbitrarily chosen numerical scale. The length of a thin column of mercury enclosed in a glass envelope is a good example. At the freezing point of pure water, the position of the top of such a column can be arbitrarily chosen as $0°$. At the boiling point of pure water, the position can be chosen as $100°$. Such a scale is called the *centigrade* or *Celsius* scale. *Any scale can be made a temperature scale.* The main consideration is: will the rest of the world use it? Three that are in use are shown in figure 4-1. Formulas for converting from one scale to another are given in the Appendix, page 304.

The most important of these from the scientific point of view is the *Kelvin* scale. It is the only scale whose individual values are proportional to the average molecular kinetic energies. For instance, at $100°$ on the Kelvin scale, the average molecular translational kinetic energy is twice that at $50°$ on the Kelvin scale. This proportionality is not true of other scales.

Specific Heat

The *specific heat* of a substance is the amount of energy needed to raise the temperature of one gram of the substance one degree centigrade.

For example, it takes 4.184 joules to raise the temperature of 1 g of water by 1 degree centigrade. Hence, its specific heat is 4.184 J/g. (There is another unit of energy called the *calorie;* 1 calorie = 4.184

KINETIC THEORY

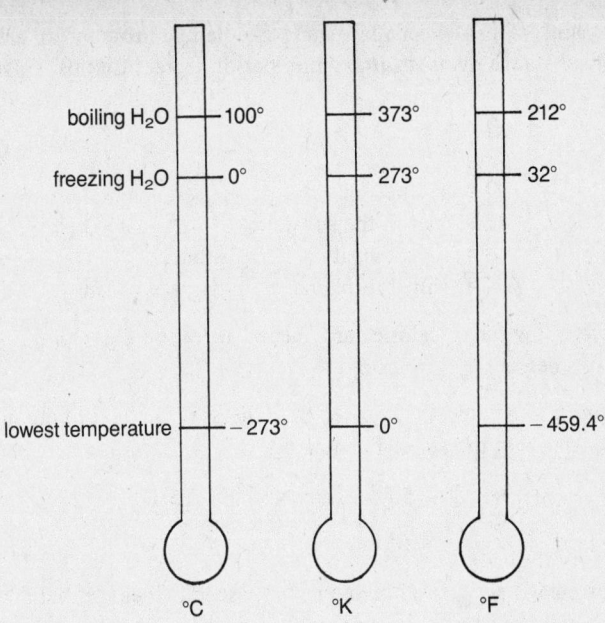

Fig. 4–1 Temperature scales

joules. Calories are often used when dealing with heat problems. In the example just noted, the specific heat of water is 1 cal/g.)

Specific heat is basically a measure of how difficult it is to change the temperature of a body. Two substances can have different specific heats because the same amount of heat flowing into each body is distributed differently among the various internal energy components. The relative temperature rise is determined by the fraction of the heat that goes into increasing the translational kinetic energy. Table 4-1 gives the specific heats of various substances.

TABLE 4-1. SPECIFIC HEATS (CALORIES/GRAM - °C)

Aluminum	.22
Copper	.093
Lead	.031
Water	1.00

The amount of heat needed to raise the temperature of an *arbitrary* mass of substance by an *arbitrary* number of degrees Celsius is given by:

$$Q = m \cdot c \cdot \Delta T$$

where: Q = heat needed
m = mass of the substance
c = specific heat of the substance
ΔT = temperature change in degrees Celsius

Example: How many calories are needed to raise 50 g of aluminum by 15 Celsius degrees?

Solution: From table 4-1, we see that the specific heat of aluminum is .22 cal. Applying the above formula, we get:

$$Q = (50g)(.22 \text{ cal/g} \cdot {}^\circ C)(15^\circ C)$$
$$= 165 \text{ cal}$$

This formula is also useful in determining specific heats experimentally. If we know how much heat is needed to raise a known mass a given number of degrees, then we can use the above formula to calculate c.

Example: Determine the specific heat of glass if 64 cal of heat can raise the temperature of 50 g of glass by 8 Celsius degrees.

Solution: Substituting these numbers in the formula gives the following expression:

$$64 \text{ cal} = 50 \text{ g} \cdot X \cdot 8^\circ C$$

We have used X to represent specific heat so as not to confuse $^\circ C$ with c. Dividing both sides by 50 g, then again by $8^\circ C$ we get:

$$\frac{64 \text{ cal}}{50g \cdot 8^\circ C} = .16 \text{ cal/gm} \cdot {}^\circ C$$

Heat Transfer

Heat can be transferred by three different methods. In *conduction*, energy is transferred molecule by molecule from one end of the material to the other. The position of the molecules does not change. The handle of a frying pan, for example, heats up by conduction. In *convection*, material is transferred from a region of high temperature to a

KINETIC THEORY

region of lower temperature. Hot air forced through vents in a heating system is an example of convection. *Radiation* is a form of heat transfer which requires no medium at all. The sun heats by radiation.

Expansion and Contraction

Most substances expand upon heating and contract upon cooling. Heating increases the kinetic energy of molecular motion; this leads to an increased average distance between the molecules. Upon cooling, the opposite happens. Expansion and contraction, of course, form the basis of practical thermometers. They are also involved in the working of the *thermostat*, illustrated in figure 4-2. Equal lengths of two different metals are welded together. When the temperature changes, the length of one changes more than the length of the other. Since they are stuck together, the only motion possible is for the pair to bend. This bending can be used to make or break a switch contact.

Two parts of the same material can heat up differently. This leads to unequal expansion and great stresses within the material. Glass, for instance, cracks under these stresses unless it is heated very slowly. The slow heating allows time for a reasonably equal heat distribution and a minimal amount of stress. *Pyrex* glass is fabricated to avoid these stresses regardless of how it is heated. That is why it can be used for cooking.

Occasionally, the generalization regarding expansion upon heating is violated. There is a form of iron which expands, contracts a little, then expands again when heated. Water also behaves anomalously. At 4°C, it expands whether it is heated or cooled. These phenomena occur because of certain rearrangements of the molecules of each substance at particular temperatures.

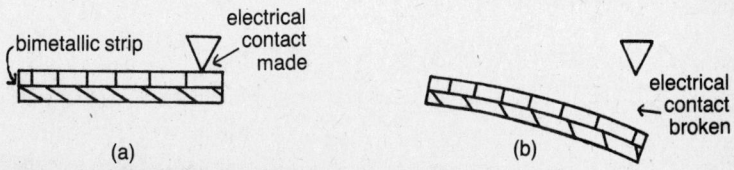

Fig. 4–2 Thermostat under two conditions: cool (*a*), hot (*b*)

Latent Heat

Latent heat is heat that flows in or out of a substance during a change of state. The heat that a solid must absorb at its melting point in order to melt into a liquid at the same temperature is called the *latent heat of fusion.* The liquid must also give up this heat in order to turn back into a solid. The heat that a liquid must absorb at its boiling point in order to vaporize is called the *latent heat of vaporization.* Similarly, the vapor must give up this heat in order to become a liquid. Latent heat flowing into a substance breaks the bonds that hold molecules together. When the change of state is the opposite way, the reforming of those bonds liberates latent heat. For water, the latent heats of fusion and vaporization are respectively, 334.7 joules/gram and 2,678 joules/gram.

IDEAL GASES

An *ideal gas* is one that has the following characteristics:

a) The randomly moving molecules exert no attractive or repulsive forces on each other except when they collide.
b) When they collide, they bounce off each other without losing energy.

Ideal gases do not exist. However, many gases exhibit behavior similar to ideal gases. Learning about ideal gases, which is easy, thus provides valuable information about real gases. For ideal gases, three quantities are readily measurable: volume, temperature, and pressure. The volume is the volume of the container; the temperature is essentially the average molecular kinetic energy. The pressure is due to the force of all the randomly moving molecules hitting the wall. In a container with several different gases, each type of molecule hits the wall and exerts a pressure independent of the other types. The pressure due to just one species of molecule is called the *partial pressure.* The pressure that we measure is just the sum of all the partial pressures.

Gas Laws

These are laws that relate the pressure, volume, and temperature:

KINETIC THEORY

1) *Boyle's Law.* For a gas at a fixed temperature, the product of pressure and volume is constant:

$$PV = \text{constant} \quad T \text{ held fixed}$$

2) *Charles' Law.* For a gas held at a fixed pressure, the quotient of volume and temperature is constant:

$$V/T = \text{constant} \quad P \text{ held fixed}$$

3) *Ideal Gas Law.* Combines the result of the above laws, holding no quantity fixed:

$$PV/T = \text{constant}$$

The *mathematical* interpretation of these formulas is simple. In an experiment governed by any of the above equations, each of the quantities on the left can vary. However, the particular *combination, PV, V/T* or *PV/T,* depending on the situation, must remain constant.

Example: A gas is contained by a piston in a 1 liter cylinder, figure 4-3, at a pressure of 2000 pa. The temperature is constant. If the piston moves down so that the new volume is 0.5 liters, what is the new pressure?

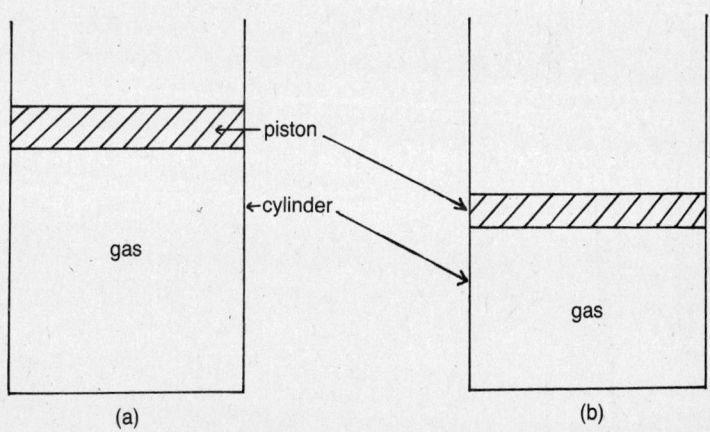

Fig. 4–3 Gas contained in a cylinder. Volume in (*b*) is smaller because the piston has been pushed down.

Solution: The product *PV* must be constant. The original measurements give:

$$(2000 \text{ pa}) \times (1 \text{ liter}) = 2000 \text{ pa} \cdot \text{liter}$$

The new measurements must give the same product:

$$P \times (0.5 \text{ liter}) = 2000 \text{ pa} \cdot \text{liter}$$

Solving for *P*, we get:

$$P = \frac{2000}{0.5} \text{ pa} = 4000 \text{ pa}$$

We halved the volume and the pressure doubled.

The *physical* interpretation of the gas laws lies in the kinetic theory. For the constant temperature case, the average translational kinetic energy, and therefore molecular speed, is constant. When the volume is increased, it takes longer for the molecules to reach the sides of the container. Hence, fewer molecules per unit time strike the walls. Also, the area of the walls increases. From the definition of pressure, $P = F/A$, we see that an increased area means a decreased pressure. Thus, when the volume is increased, the pressure goes down.

For the constant pressure case, an increase in temperature results in increased molecular speeds. The molecules hit the sides of the container with increased force, tending to raise the pressure. Since the pressure must remain constant, the piston must rise to relieve the excess pressure. Hence, as the temperature increases, the volume increases.

Evaporation and Boiling

When a liquid evaporates, molecules at the surface that are moving fast enough, and in the upward direction, can escape. These escaping molecules give rise to a small pressure called the *vapor pressure.* Evaporation can be speeded up by heating the liquid. This causes all the molecules to move faster, increasing the number and energy of those escaping. Clearly, this increases the vapor pressure. How high can the vapor pressure go? Only as high as the outside pressure. When the vapor pressure reaches the outside pressure the liquid is said to *boil.* Note that the criterion for boiling does not involve temperature. When we say that the boiling point of water is 100°C, we mean that water at sea level must be heated to 100°C in order that its vapor pressure equal the

KINETIC THEORY

atmospheric pressure. On Mount Everest, the atmospheric pressure is lower than at sea level. Hence, we don't have to bring the water to as high a temperature to raise the vapor pressure to the required amount. Thus, the boiling point is less than 100°C. Inside a pressure cooker, the trapped air is at a pressure equal to twice the normal atmospheric pressure. The water must be heated to higher than 100°C to raise the vapor pressure high enough for boiling.

THE LAWS OF THERMODYNAMICS

Thermodynamics is essentially the study of heat and related phenomena in terms of readily observable variables. Examples of such variables are pressure, temperature, and volume. The speed of gas molecules, for instance, is not observable. In fact, thermodynamics makes no assumptions about the behavior or even the existence of molecules. Many concepts of thermodynamics are subtle and elusive, so we will concentrate on a few important ideas.

Heat Engines and Refrigerators

We already know that heat flow between two bodies at different temperatures is from hot to cold. For convenience, we will call these bodies *hot* and *cold reservoirs*. A *heat engine* is a device that takes some heat from this flow and converts it into work. A *refrigerator,* on the other hand, takes work and uses it to transfer heat from a cold to a hot reservoir. These devices are shown schematically in figure 4-4.

First Law of Thermodynamics

This law says that *energy can be neither created nor destroyed*. Energy may change from one form to another, but the total amount in the universe is constant. If this is true, why do people talk of an *energy shortage?* To answer this, we must first look at another law.

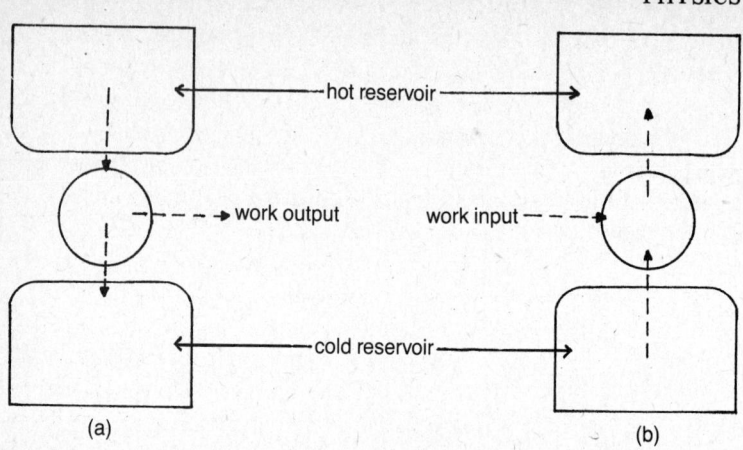

Fig. 4–4 Heat engine (*a*) and refrigerator (*b*). Vertical arrows show heat flow.

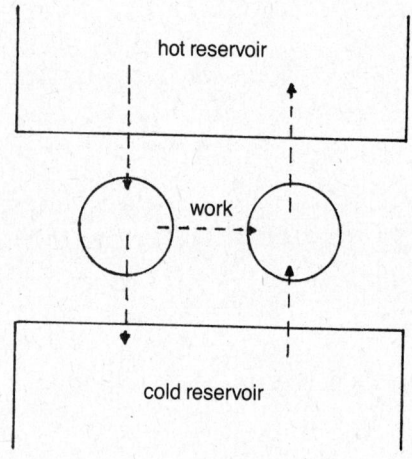

Fig. 4–5 Theoretical coupling of heat engine and refrigerator between the same reservoirs

KINETIC THEORY

Second Law of Thermodynamics

The *Second Law of Thermodynamics* says essentially that there can be no transformation of energy without some of the energy being lost to unrecoverable heat.

In order to understand this more clearly, we will assume that we can violate the second law. To do this, we will construct a device which is a heat engine and a refrigerator coupled as in figure 4-5. We know that heat normally flows from the hot reservoir to the cold until there is no longer a difference in temperature. The engine takes some of this heat and converts it to work. Now, imagine trying to keep the engine working forever; this requires that the original temperature difference between hot and cold reservoirs be maintained. This is where the refrigerator comes in. It uses the work done by the engine and delivers to the hot reservoir all the heat that flows to the cold reservoir. This restores the original temperature difference, enabling this process to go on forever.

The second law forbids the above scenario. When the engine operates, heat is used to do work. When the refrigerator operates, work is used to transfer heat. In these two energy transformations, some heat must be dissipated. The dissipated heat is spread evenly throughout the system. Because of this dissipation, the amount of heat we have left to deliver back to the hot reservoir is *less* than what we need to maintain the original temperature difference. Eventually, no temperature difference exists and the engine has to stop. Note that the total amount of energy in the system remains constant; the first law is not violated. The problem is that eventually, because of dissipation, all the energy will be spread evenly over the system. The temperature of the system will be uniform and there will no longer be any hot and cold reservoirs, which are needed to make engines work.

What holds true for our little system holds true for the universe. Eventually, all the energy of the universe will be spread evenly throughout. There will be no hot or cold reservoirs and no way for heat engines to work.

5
Electricity and Magnetism

The key to electric and magnetic behavior is the concept of *electric charge*. We have all experienced the phenomenon of electric charge. If you walk across a rug on a dry day and touch a doorknob or other metal object, you will get a shock and hear a crackling noise. Look carefully when the lights are out and you will notice a spark. If you rub a comb through your hair under similar dry conditions, it will attract small pieces of paper. In these examples, your body and the comb are said to possess electric charge.

STATIC ELECTRICITY

Static electricity deals with electric charges that are at rest. They behave differently from charges in motion.

Types of Charge

Experiments have shown that there are only two different types of charge. These have *arbitrarily* been called *positive* and *negative* charge, symbolized by $+Q$ and $-Q$. It is irrelevant which is called which. What is important is the following rule which summarizes the results of numerous experiments concerning charges:

Like charges repel; unlike charges attract.

The standard unit of electric charge is called *coulomb* (C), which will be defined precisely in the next section.

ELECTRICITY AND MAGNETISM

Coulomb's Law

Coulomb's Law relates the force, f, between two charges to the amount of charge, Q_1 and Q_2, and the distance r, separating them:

$$f = kQ_1Q_2/r^2$$

The constant k has the value 9×10^9 Nm^2/c^2, provided the charge is measured in coulombs and the distance is measured in meters. There are several important ideas stemming from Coulomb's Law:

1) The *direction* of f is along the line joining the charges.
2) A *coulomb* is a quantity of charge such that when two charges are placed one meter apart, the force between them is 9×10^9 N.
3) The force is *negative* if the charges are of *opposite* sign, *positive* if they are of the *same* sign. Since opposite charges attract, we interpret a negative force as attractive. Similarly, a positive force is interpreted as repulsive.
4) The *form* of Coulomb's Law is the same as the form of Newton's Law of Gravitation, with charge playing a role analogous to mass. However, all gravitational forces are attractive whereas not all electric forces are.
5) The *strength* of the electric force is much greater than the strength of the gravitational force. The gravitational force between two masses of one unit each placed one meter apart is much less than the electric force between two one-unit charges placed the same distance apart. To see this, just put mass, charge, and distance equal to 1 in the two formulas. Then we get $F_{grav} = 6.67 \times 10^{-11}$ N and $F_{elec} = 9 \times 10^9$ N. In practical terms, enormous masses are needed in order to exert the same force as two one-coulomb charges.

Electric Field

The *electric field*, E, can be described by a mathematical expression that gives the force on a one-coulomb positive charge at any given point in space. Consider the simple case of two charges, Q_1 and Q_2. We can say that because of its charge, Q_2 somehow alters the space around itself. We speak of the space around Q_2 as being altered by an electric field, E. Then the force on Q_1 can be thought of as an interaction between Q_1 and the electric field. The calculation of the force on Q_1 then depends only on a knowledge of the field, not the charge that produced it. Since E gives the force on one coulomb, the force on Q_1 coulombs is just $E \times Q_1$.

This may seem like a roundabout way of calculating the force on one charge due to another; we could have simply used Coulomb's Law. However, we are often confronted with a complicated distribution of charges. It turns out that for such charge distributions, there are powerful mathematical tools available for figuring out E. The force on a charge Q in the vicinity of such a charge distribution is just $Q \times E$. This is much easier than calculating the Coulomb force for many charges and getting the vector sum of the resulting forces.

The electric field is conveniently represented in diagrams by *lines of force*. These are just lines drawn to show the direction of the field at many different points. Figure 5-1 shows the lines of force around positive and negative charges.

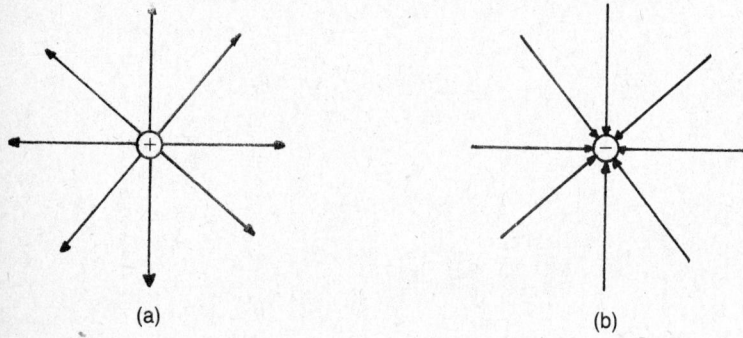

Fig. 5–1 Electric field around positive (*a*) and negative (*b*) charge

Potential Difference

The *potential difference* between two points in an electric field is the work done in moving a one-coulomb charge from one point to the other. This is analogous to the difference in gravitational potential energy between two points, which is the work done in lifting an object from the lower point to the higher one.

Potential difference has the units of work per unit charge, joules per coulomb. One joule per coulomb is called a *volt*.

ELECTRICITY AND MAGNETISM

ELECTRIC CURRENT

An *electric current* exists when charges are in motion. An important example of this is the motion of negatively charged particles called *electrons*. Certain materials, called *conductors*, have electrons which can move freely through the material in the presence of an electric field. Metals are notable examples of conductors. Materials which do not conduct are called *insulators*. Plastics and glass are insulators. There are also materials that conduct under certain conditions but do not conduct under others. They are called *semiconductors*. The two most common semiconductors are germanium and silicon.

Batteries and Direct Current

A *battery* is a device which maintains an excess of electrons at one terminal and a deficiency of electrons at the other. These are called, respectively, the negative and positive terminals. Because of this imbalance, a potential difference exists between the terminals. Thus, the battery can do work on electrons. When a wire is connected between the terminals, electrons which are in excess at the negative terminal are pushed through the wire to the positive one. The chemicals in the battery maintain the excess at the negative terminal and thus maintain the flow. This steady flow is called a *direct current*. One *ampere* of current is the flow of 6.25×10^{18} electrons past a given point per second; this is the same as one coulomb per second.

Electric Circuit

A *closed circuit* is one which provides an uninterrupted path for electron travel from negative to positive terminal. An *open circuit* is one where the path is broken. Current does not flow in an open circuit. The *resistance* of a circuit is its opposition to the flow of current and is measured in *ohms* (Ω). One ohm is the resistance of a circuit in which one ampere of current flows as a result of a potential difference of one volt. The interconnecting wires of circuits have very little resistance. Usually, it is desired that a particular circuit have a specified resistance. This is done by inserting *resistors*, which are devices that provide the desired values.

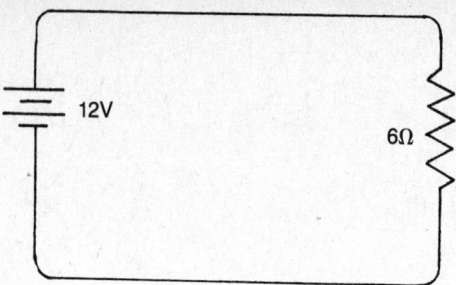

Fig. 5–2 Simple circuit

Circuit Relationships

A simple circuit is shown in figure 5-2. It consists of a 6 ohm resistor and a 12 volt battery.

1) *Ohm's Law* relates the resistance (R), voltage (V), and current (I):

$$V = I \times R$$

For this circuit, the current flowing is:

$$12 \text{ volts} = I \times 6 \text{ ohms}$$

Therefore:
$$I = 12 \text{ volts} / 6 \text{ ohms}$$
$$I = 2 \text{ amps}$$

2) *Power dissipation* occurs in the resistor according to the relation:

$$P = I^2 R$$

Since $R = V/I$, this can also be expressed as:

$$P = VI$$

We can verify that this actually has the units of power:

$$P = \text{volts} \times \text{amps}$$
$$= (\text{joules/coulomb}) \times (\text{coulombs/second})$$
$$= \text{joules/second} = \text{watts}$$

The power dissipated in the above resistor can be figured out with either formula:

$$P = (2 \text{ amps})^2 \times (6 \text{ ohms}) = 24 \text{ watts}$$
$$\text{or}$$
$$P = 2 \text{ amps} \times 12 \text{ volts} = 24 \text{ watts}$$

ELECTRICITY AND MAGNETISM

Dissipated power appears as heat in the resistor.

3) A *series circuit* results when the circuit of figure 5-2 is modified by the addition of resistor R_1, as in figure 5-3a. The two resistors here are said to be in *series*. They can be replaced by one whose value, R_t, is:

$$R_t = R_1 + R$$

If $R_1 = 6$ ohms, then $R_t = 6 + 6 = 12$ ohms. The new resistance is greater than the original because a new obstacle has been placed in the path of the electrons.

4) A *parallel circuit* results when R_2 is added to the circuit of figure 5-2, as shown in figure 5-3b. The resistors here are said to be in *parallel*. They can be replaced by one whose resistance, R_p, is given by the following formula:

$$1/R_p = 1/R + 1/R_2$$

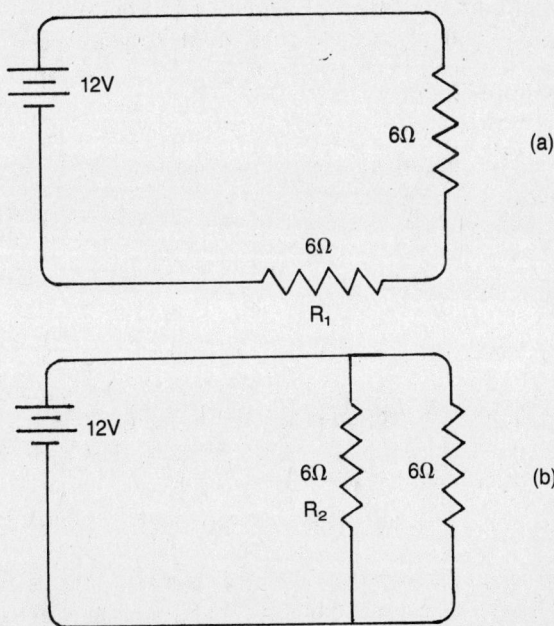

Fig. 5-3 Series (*a*) and parallel (*b*) circuits

In our example, the new resistance is:

$$1/R_p = 1/6 + 1/6$$
$$1/R_p = 1/3$$
$$R_p = 3 \text{ ohms}$$

The new resistance is less than the original because an alternate path has been provided for the electrons.

A summary of electrical quantities is given in table 5-1.

TABLE 5-1. SUMMARY OF ELECTRICAL QUANTITIES

Electric field	newtons per coulomb
Potential difference	joules per coulomb = volts
Current	coulombs per second = amperes
Resistance	volts per ampere = ohms

MAGNETISM

We are all familiar with the action of a magnet, having seen one attract pieces of iron or deflect the needle of a compass. We can make an analogy with the electric case and say that the magnet somehow alters the space around it, creating a *magnetic field*. An appropriate object in that space feels a force. In the electric case, the appropriate object is a charge; in the magnetic case, an appropriate object is, say, a piece of iron. The underlying cause of all magnetic fields is electric charge in motion.

Magnets

1) A *magnetic moment*, figure 5-4, is a magnetic field due to a charge moving in a circle. Inside atoms are electrons which move in circular paths. Thus, each electron gives rise to a microscopic magnetic field. If two electrons move in opposite directions, then their magnetic moments cancel out. In a macroscopic piece of material with a huge number of atoms, the randomly oriented magnetic moments generally cancel out.

2) A *magnetic domain*, figure 5-5, is a tiny but distinct region of material whose magnetic moments are lined up in essentially the same direction. Although there are about 10^{20} atoms in a domain, their

ELECTRICITY AND MAGNETISM 57

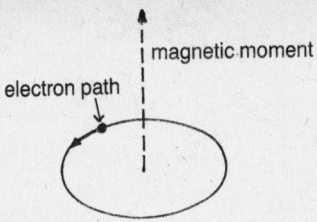

Fig. 5-4 Magnetic moment of electron moving in a circle

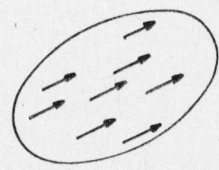

Fig. 5-5 Single magnetic domain. Arrows represent magnetic moments of individual atoms.

magnetic moments do *not* cancel. They combine in such a way that the domain acts like a magnet. Materials with these domains are called *ferromagnetic*. Normally, these domains are aligned in random fashion (figure 5-6). However, when an external magnetic field is present, the domains line up in the same direction as the external field (figure 5-7). When the external field is removed, the domains return to their random orientation.

3) A *permanent magnet* is a material whose domains are permanently lined up in the same direction, as in figure 5-7. When suspended freely, the magnet is oriented so that one end points to the earth's north pole and the other points to the south pole. We call these the *north* and *south* poles, respectively. From experimental evidence, we have the following rule:

Like poles repel; unlike poles attract.

This looks like the rule for electric charges. However, although single charges exists, single magnetic poles do not. If we cut a bar magnet in

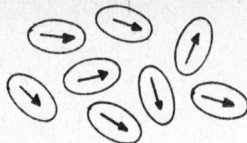

Fig. 5-6   Randomly oriented magnetic domains

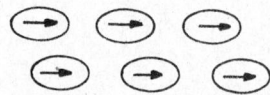

Fig. 5-7   Uniformly oriented magnetic domains

half, the north and south poles are not separated; each part still has its own north and south pole. The north-south designation serves as a convenient way to denote the direction of the permanently aligned magnetic domains. This direction is from south to north. Cutting the magnet in half does not change the orientation of those domains. Each piece still has permanently aligned domains and like the original piece will seek the north and south directions.

Production of Fields by Currents

Since moving charges produce magnetic fields, it is natural to investigate the field due to different current configurations. The idea of lines of force, used to show the direction of electric fields, can be used exactly the same way for magnetic fields.

1) The *field around a long, straight wire* is tangent to a circle that is concentric with the wire, as shown in figure 5-8. In this diagram, the wire is perpendicular to the page so we just see its cross section.

2) The *field on the axis of a loop of wire* is seen in figure 5-9.

3) The *field inside a solenoid* is seen in figure 5-10. A solenoid is a coil of many turns wrapped around a cylinder. The magnetic field is stronger than for a single coil simply because there are many more loops of wire. It can be made even stronger by inserting a piece of iron

ELECTRICITY AND MAGNETISM

Fig. 5–8 Field around a long straight wire. The magnetic field at any point is tangent to a circle centered at the wire and passing through the point.

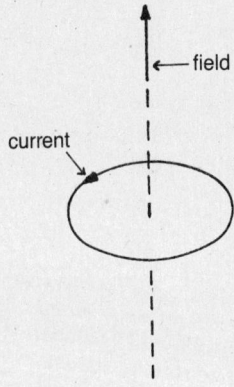

Fig. 5–9 Field on the axis of a loop of wire

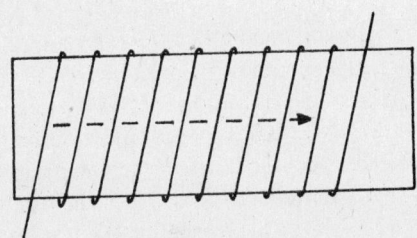

Fig. 5–10 Solenoid field is parallel to the axis of the solenoid.

inside the coil. This is the basis for the *electromagnet,* a device which is widely used in industrial and consumer products.

Magnetic Fields and Forces

Experiments show that a charge moving *across* magnetic field lines experiences a force. The field, the motion of the charge, and the force form three mutually perpendicular directions as shown in the vector diagram below:

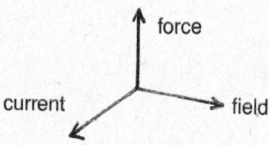

Assume for the following examples that the field is perpendicular to this page.

1) A charge moving in the plane of this page always cuts across lines of force. It is continuously deflected perpendicular to its direction of motion and ends up going in a circle.

2) An electron in a wire can move only along the wire. A current-carrying wire in the plane of this page will be deflected perpendicular to its length. However, if the wire is oriented so that it is perpendicular to this page, i.e., in the direction of the field, then it does not cut magnetic field lines and experiences no force.

Generation of Current by Magnetic Fields

The results of many experiments show that magnetic fields can be used to generate currents.

1) A *wire moving with respect to a field* generates a current, as long as it moves across magnetic lines of force. It does not matter whether the wire moves or the field moves, as long as there is relative motion.

2) A *changing magnetic field* can generate a current. If a stationary loop of wire is placed across the lines of force of a magnetic field whose strength is changing, a current is generated in the wire. However, if the field is constant, no current will be generated.

# ELECTRICITY AND MAGNETISM

Applications

1) An *alternating current generator* is basically a loop of wire moving through a magnetic field (figures 5-11 and 5-12). For convenience, the part of the loop next to the south pole will be called the south side; the one next to the north pole, the north side. The loop is turned by an outside force and a current is generated. Assume the loop is turned continuously in the same direction. Then the current in the loop must be such that its direction in the south side is always one way and its direction in the north side is always the opposite way. Since sides A and B of the loop switch places every half turn, we can see that the circulation of current in the loop must reverse itself accordingly. The current output will be one way for half a turn and the other way for the other half turn. This is called *alternating current* (AC). We call the potential difference at the output of the generator the *electromotive force*.

2) An *electric motor* is shown in figure 5-13, highly simplified. It is basically a current-carrying wire loop in a magnetic field. In our example, current is flowing in the loop from B to A. Thus, there is a force on the south side of the loop perpendicular to both the wire and the field. By our convention, it is out of the paper. The force on the north side is into the paper because the current is in the opposite direction. These two forces produce a torque which tends to spin the loop. We would like the loop to spin continuously in the same direction. Thus, the forces on the south and north sides must always be in the directions just described. This means the current on the south side must *always* be downward; on the north side, *always* upward. Note, however, that after half a turn B is near the south pole and A is near the north pole, as shown in figure 5-14. If the current continues to flow from B to A, then the direction of current in the south and north sides of the loop will be opposite what they are supposed to be. Thus, an arrangement must be made to allow the current to reverse its circulation in the loop every half turn. This happens automatically if the motor works off AC. However, it it works off DC, a *commutator*, shown in figure 5-15, is needed. One of the half rings is connected to the positive terminal, the other to the negative terminal. As the loop spins, the leads alternately contact the positive and negative terminals of the power source, causing the current flow in the loop to reverse accordingly.

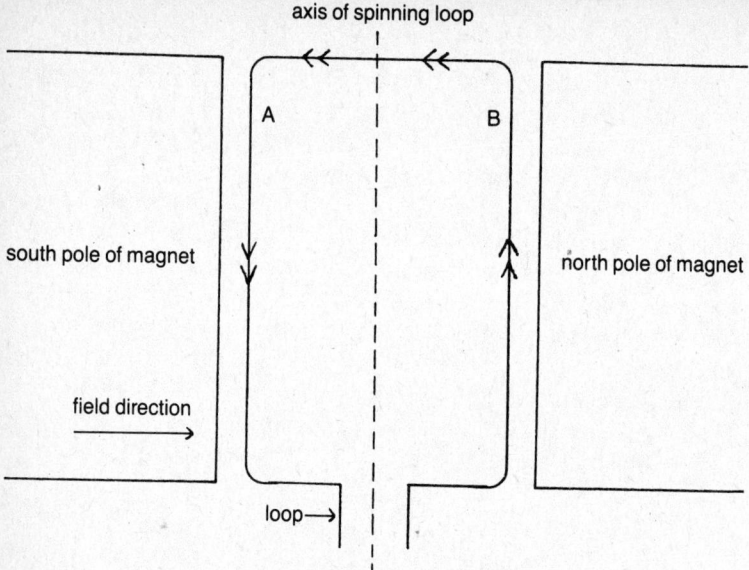

Fig. 5–11 Spinning loop in a magnetic field. As the loop spins, sides *A* and *B* cut magnetic lines of force, so a current is generated.

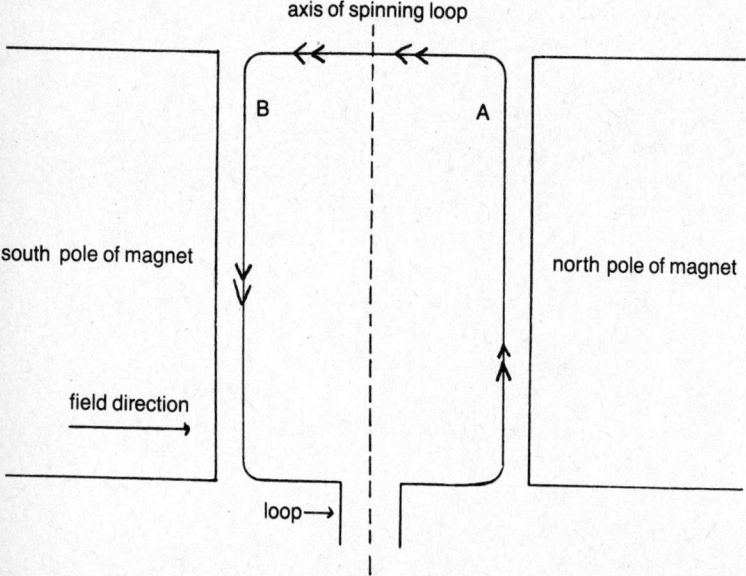

Fig. 5–12 Same loop half a turn later. Note that the current in the loop has reversed (A→B, not B→A).

ELECTRICITY AND MAGNETISM

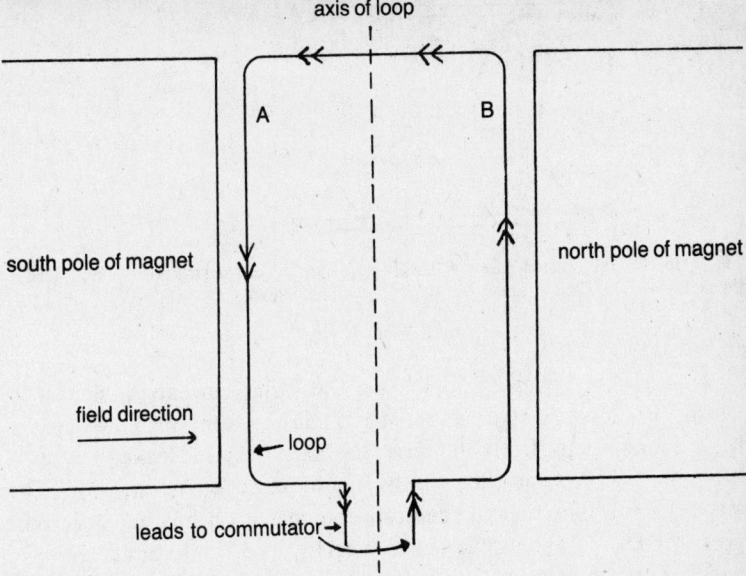

Fig. 5–13 Simple electric motor. Current (shown by double arrows) applied to the loop causes it to spin.

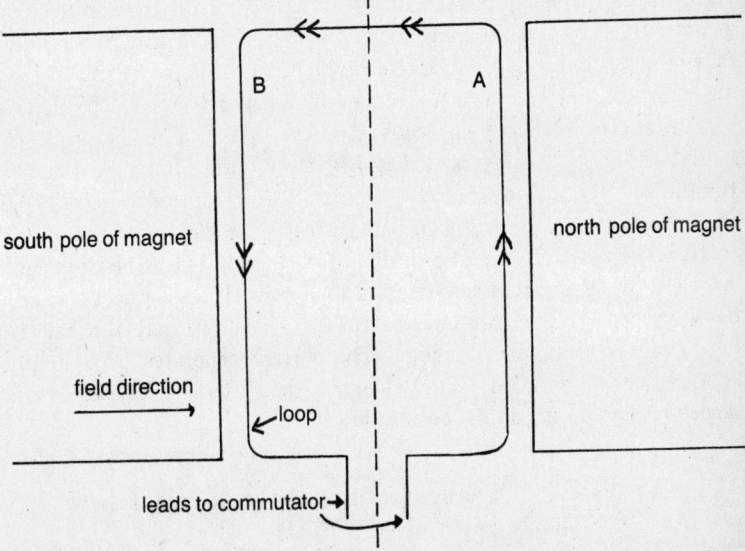

Fig. 5–14 Same loop half a turn later. If the loop is to spin the same direction, current applied to it must reverse directions every half turn. Note that current flows from $A \rightarrow B$ here, not $B \rightarrow A$.

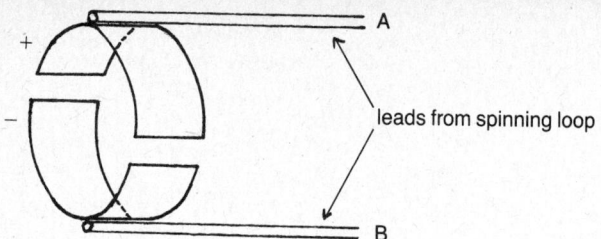

Fig. 5-15 A commutator. After one half turn, lead B will contact the positive ring, lead A the negative ring. The current in the loop will then reverse.

3) A *galvanometer* is a device for measuring currents. A simplified version is shown in figure 5-16. It is similar to a motor; however, the loop cannot spin freely because the axis is spring-loaded to resist motion. Also, current flows only in one direction in the loop. The amount of turning that the coil can do depends on how much current goes into the coil. An indicator connected to the coil points to a scale which is calibrated to read current.

4) A *transformer* is a device that changes AC voltages. It is an application of current generation by a changing magnetic field. A simplified model is shown in figure 5-17. If a battery is connected to coil A, then a magnetic field due to that current will develop. During the time it takes for the field to develop, coil B will be in the presence of a changing magnetic field and will have a current generated in it. When the current in coil A becomes steady, no current is generated in coil B. If the current in coil A is turned off, then during the time it takes for the magnetic field of A to disappear, coil B will once more be in the presence of a changing magnetic field. The current generated in B, however, will be opposite in direction to what it was before, since the field of A is decreasing instead of increasing. If we connect a source of AC to coil A, then we have a situation similar to that of a battery being turned on and off rapidly. The current generated in B simply follows the alternations of the changing field of A. Thus, we have a device for generating an AC current in coil B.

There is a simple relation which gives the voltage developed in coil B:

$$\text{voltage at } B = (\text{turns of } B/\text{turns of } A) \times \text{voltage at } A$$

ELECTRICITY AND MAGNETISM 65

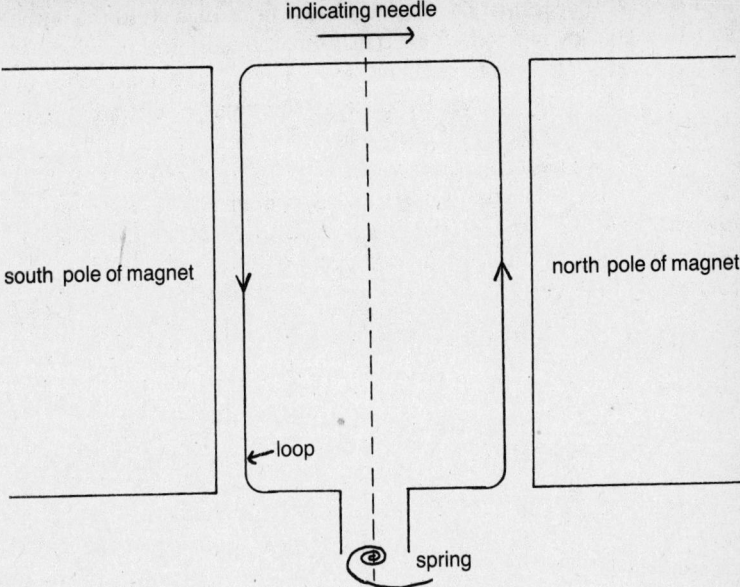

Fig. 5-16 Galvanometer. A current in the loop causes the loop to twist against the resistance of the spring.

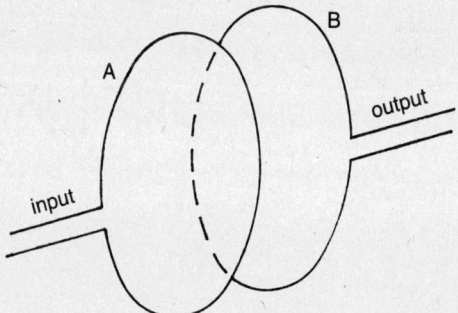

Fig. 5-17 A very simple transformer. A change in the magnetic field produced by A induces a current in B.

In a *step-up* transformer, B has many more turns than A, so the output voltage is greater. In a *step-down* transformer, B has fewer turns than A, so the output voltage is lower. Since we can never have more power out than we put in, an increased voltage at the input means a decreased current at the output. This principle is very important in the transmission of power over long distances. In order to minimize the dissipation of power, I^2R, the voltage is stepped up before transmission, lowering the current and, therefore, the I^2R losses. Just before the power lines reach a house, the voltage is stepped down to its normal value.

6
Wave Motion

A periodic disturbance is one which repeats itself at regular intervals. A *wave* is a periodic disturbance which can propagate, usually through a medium, but sometimes through empty space.

WAVE TYPES

Waves are classified according to the relation between the periodic motion and the direction of propagation. The nature of wave motion is most easily understood by considering waves as they exist in a medium. In general, waves propagate in a medium because a disturbance to one molecule is transmitted to its neighbor, and so on from molecule to molecule.

Transverse Waves

In a *transverse wave*, the motion of the disturbed molecule is *perpendicular* to the propagation of the wave. A simple example is shown in figure 6-1. One end of the rope is attached to a wall and the other end is wiggled up and down. Each molecule of the rope oscillates up and down. However, a given molecule begins its oscillation slightly later than the one before it. The wiggle, but not the molecules, moves to the right.

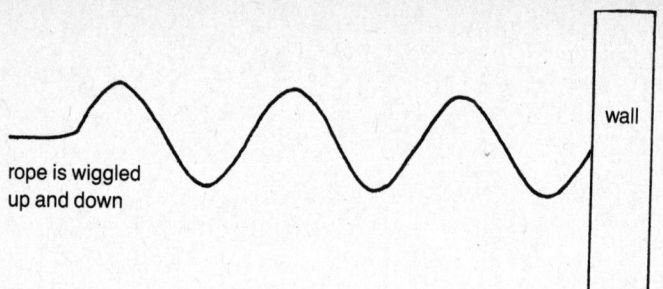

Fig. 6–1 Transverse wave in a rope

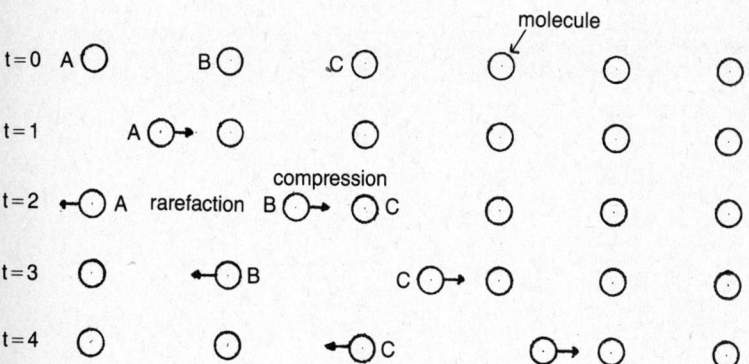

Fig. 6–2 Longitudinal wave proceeding down a line of molecules, shown here for five instants of time. Arrows indicate how the molecule moves at each instant.

WAVE MOTION

Longitudinal Waves

Longitudinal waves are waves in which the motion of the disturbed molecule is *along* the direction of propagation. Figure 6-2 shows a *highly simplified* model of a longitudinal wave produced in a line of molecules initially at rest. Imagine that something has caused A to move slightly to the right. When A moves to the right, it pushes B to the right; B then pushes C, and so on down the line. The bunching up of molecules is called a *compression;* the space that results is called a *rarefaction*. One can see from the diagram that both the compression and the rarefaction travel down the line of molecules, even though the molecules themselves do not go anywhere.

Standing Waves

A *standing wave* is an oscillatory motion which does not propagate at all. When a violin string is bowed, transverse waves move rapidly from one end to the other, getting reflected at the fixed boundaries. The result of these reflections is the development of certain equally spaced points on the string, called *nodes*, which do not move at all. Once the nodes are established, the disturbance no longer travels along the string. All the points between two nodes simply oscillate up and down in phase, but with different amplitudes. Figure 6-3 shows a standing wave with three nodes, including the endpoints. Points A and B are called *antinodes*.

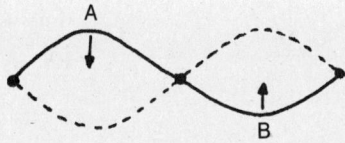

Fig. 6–3 Standing wave with three nodes. Arrows indicate direction of motion at this instant.

WAVE CHARACTERISTICS

Both longitudinal and transverse waves can be represented as shown in figure 6-4. Here, the displacement of a particular molecule, whether longitudinal or transverse, is indicated by the vertical coordinate. The tops of the peaks are called *crests*. The bottoms of the valleys are called *troughs*. The height of the crests, which is the same as the depth of the troughs, is called the *amplitude* of the wave. Amplitude is basically a measure of how strong a disturbance is. The distance between crests is called the *wavelength*. The number of oscillations per second is called the *frequency*. This is a measure of how rapidly the disturbance vibrates. It should not be confused with *speed*, which tells how far the disturbance propagates in a given time. However, there is a relation between them:

$$\text{speed} = \text{frequency} \times \text{wavelength}$$

The *phase* is a term which describes where any given point on a wave is in its cycle. Two points which are at the same part of their cycle are said to be *in phase*. A and B in figure 6-4 are in phase. Two points which are half a cycle apart, such as A and C, are said to be *completely out of phase*.

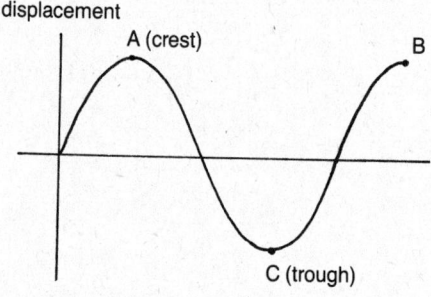

Fig. 6–4 Graph of a wave

WAVE MOTION

Fig. 6-5 Individual molecules at the surface of water execute circular motion. Dashed line shows how positions of molecules form a wave.

EXAMPLES OF WAVES

Water Waves

A wave traveling across the surface of a pond looks like the transverse wave of our rope example. There is a slight difference, however. Instead of simply oscillating up and down, the molecules of water oscillate in small circles (figure 6-5). Each circular motion begins slightly later than the one before it, so the net result is a disturbance similar to the one moving across the rope.

Sound Waves

Sound is an example of a longitudinal wave. Sound usually consists of periodic compressions and rarefactions produced by a vibrating object such as a loudspeaker. Certain explosive sounds, however, such as a firecracker or thunderclap, are not periodic; they are single compressions. Each type travels in air at a speed of about 335 m/s, although the speed increases slightly with temperature. In water, sound travels at about 1400 m/s.

The human ear can perceive sound from about 20 to 20,000 hertz (Hz) (1 Hz = 1 vibration per second). Most musical sounds are well within these limits. It is the frequency of a sound wave that determines its *pitch*. When a single note is played on an instrument, multiples of the note's frequency also are produced. The lowest frequency is called the *fundamental;* the multiples are called *harmonics*. We know that a violin and piano do not sound alike even when playing the same fundamental. This is because each instrument produces a unique *mixture of harmonics*. The ear recognizes the different mixtures as belonging to different instruments.

Electromagnetic Waves

An *electromagnetic wave* consists of periodically varying electric and magnetic fields at right angles to each other. These fields are also perpendicular to the direction of propagation, so electromagnetic waves are transverse. Since electric and magnetic fields can exist in empty space, electromagnetic waves do *not* need a medium in which to propagate.

An electromagnetic wave is produced by an oscillating electric charge. We have seen that a moving charge can produce a magnetic field. We have also seen how a changing magnetic field can produce a current. Actually, these examples are specific cases of a more general principle: a changing magnetic field produces an electric field and a changing electric field produces a magnetic field. Thus, a charge which is in oscillatory motion can create a periodic electric and magnetic disturbance.

WAVE BEHAVIOR

Point and Extended Sources

A *point source* is a point from which a wave originates. Waves spread out spherically from a point source, as indicated by the concentric circles of figure 6-6. These are called *wavefronts*. A point source is only an idealization. Real waves are produced from sources that are larger than points. These are called *extended sources*. A very simple extended source is shown in figure 6-7. It is just a line of point sources very close together. As we might expect, the wavefronts from this source are parallel to the original source. However, with some difficult mathematical analysis, it can be shown that these wavefronts are not of uniform width. The beam of parallel lines spreads out as shown in the diagram. The spread shown here is quite exaggerated, however.

Reflection

When a wave strikes an object at an oblique angle, its direction is changed. This is called *reflection*. Figure 6-8 shows how a plane wavefront is reflected from a flat surface. Each part of the wavefront is

WAVE MOTION

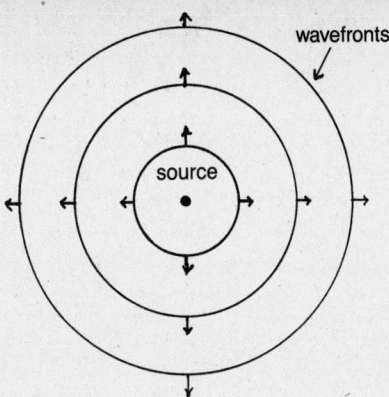

Fig. 6–6 Point source. Wavefronts travel radially outward.

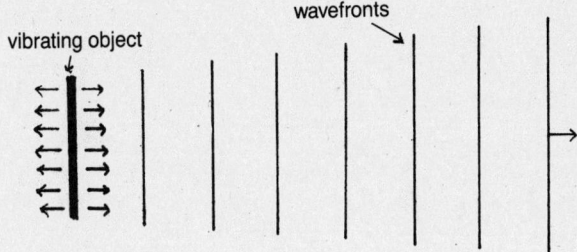

Fig. 6–7 Wavefronts produced by a flat surface

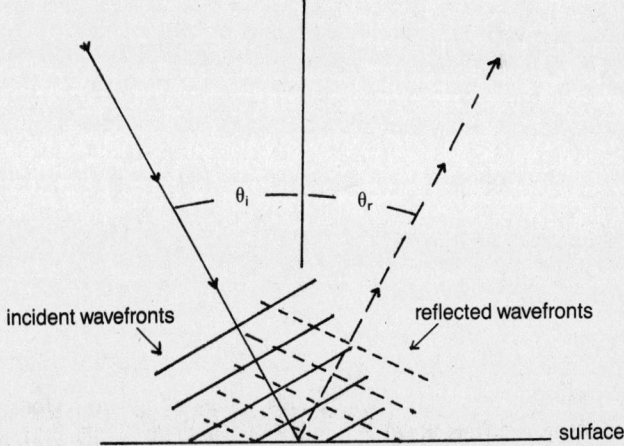

Fig. 6–8 Reflection at a plane surface

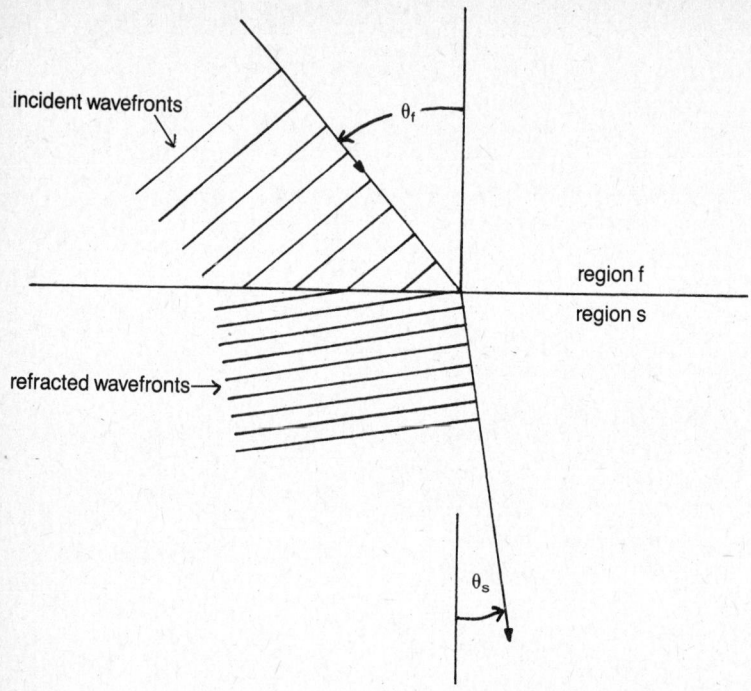

Fig. 6–9 Refraction at a plane surface. The wave travels faster in region f than in region s.

successively reflected so that its angle of incidence, θ_i, is equal to its angle of reflection, θ_r. The entire wave thus appears recreated in the new direction.

Refraction

When a wave obliquely crosses a boundary separating two regions where the wave speed differs, its direction is changed. This is called *refraction*. Figure 6-9 shows a plane wavefront about to go from a "fast" to a "slow" region. When the wavefront is partly in one region and partly in the other, the two parts move at different speeds. This causes an

WAVE MOTION

abrupt bend at the boundary, as seen in the diagram. Note that the angle in the "fast" region is greater than the angle in the "slow" region.

Interference

The combination of two waves simultaneously incident on the same point is called *interference*. Analyzing interference is complicated; we will consider two very special situations where the wavelength of the two waves is the same.

1) *Constructive interference* occurs when the two waves arrive in phase. In figure 6-10, the two incoming waves are represented by graphs A and B. The graph of the resultant wave is obtained by adding the vertical coordinates of A and B for each point along the horizontal axis. Since A and B are identical, the resultant graph has vertical coor-

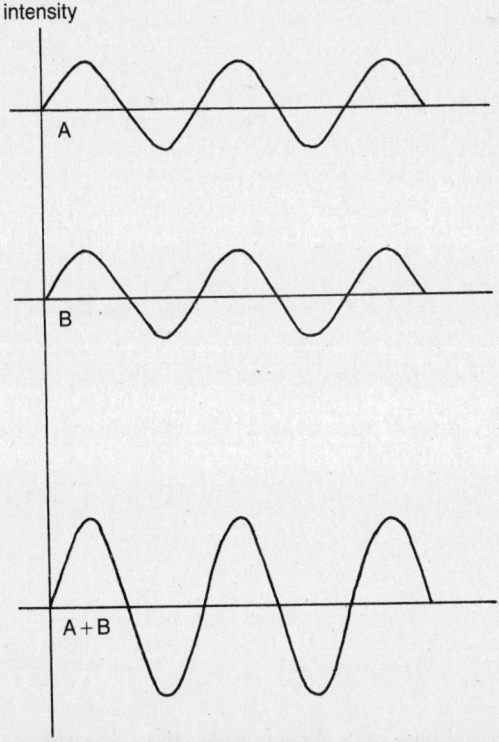

Fig. 6–10 Constructive interference. Waves A and B, arriving simultaneously at the same point, interfere to produce a wave with twice the amplitude.

dinates which are twice as high as the original. Hence, constructive interference results in a wave that has twice the amplitude of the original.

2) *Destructive interference* results when the waves arrive completely out of phase. This is seen in figure 6-11. Using the same procedure as above for adding the waves, we see that every vertical coordinate of A is matched with an equal negative one from B. The sum of the vertical coordinates is therefore zero for the entire wave. Destructive interference results in no wave at all.

Diffraction

Diffraction refers to the ability of a wave to bend slightly around the edges of obstacles. Figure 6-12 shows a wavefront incident on an ob-

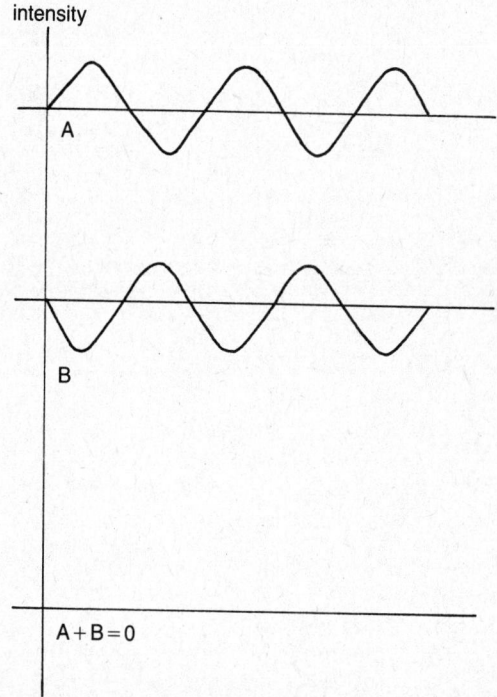

Fig. 6–11 Destructive interference. Waves A and B, arriving simultaneously at the same point, cancel each other out.

WAVE MOTION

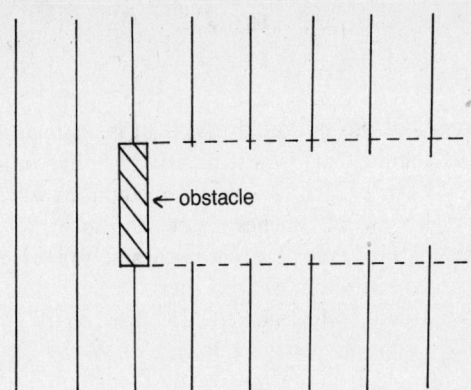

Fig. 6–12 Diffraction of waves around an obstacle. Note that wavefronts spread slightly into shadow zone.

stacle. The wave is clearly cut into two fronts. If these two fronts were to travel in a perfectly straight line, we would expect the zone between the dashed lines, called the *shadow zone,* to be free of wave activity. From our discussion of extended sources, we know that these wavefronts spread out a bit. Hence, some of the original wave enters the shadow zone. In effect, the wave has bent around the obstacle.

Doppler Effect

The *Doppler effect* is the apparent change in frequency of a wave due to relative motion between the source and an observer. It is responsible, for instance, for the change in pitch of a train whistle when a train goes by. A certain number of waves spans the distance between the source and the observer. If the relative motion of source and observer is toward each other, then the same number of waves gets squeezed into a smaller space. Hence, the wavelength gets smaller. Since smaller wavelength means higher frequency, the observer perceives a higher frequency than the source actually puts out. By similar reasoning, we can see that as source and observer move apart, the perceived frequency becomes lower.

LIGHT

In the early days of physics, there were two competing theories of light. The *corpuscular theory* stated that light is a stream of particles. The *wave theory* stated that light is a wave. At that time, it was assumed (incorrectly) that all waves need a medium in which to propagate. Since light obviously reaches the earth from the sun, it was assumed that "empty" space is filled with a substance called *ether*. Light was regarded as vibrations of this ether.

Work done by Young, who showed that light could be diffracted, and Maxwell, who showed that light is an electromagnetic wave, convinced most physicists of the validity of the wave theory. The corpuscular theory fell into disfavor. In this section, we will examine some characteristics of light.

Color

Our eyes perceive different wavelengths of light as different colors. Red and orange are long wavelengths; blue and violet are short wavelengths. *White light* is a combination of wavelengths in the proportion found in sunlight. Objects appear colored because they reflect wavelengths corresponding to those colors. If sunlight hits an object that reflects red, we will see the object as red. However, if only green light hits that object, we will see nothing, since there is no red in the incident light to be reflected.

When a band of energy, such as sunlight, is separated according to wavelength, the result is called a *spectrum*. There are basically two kinds of spectra:

1) A *continuous spectrum* is one where all wavelengths are present, with one imperceptibly blending into the next.
2) A *line spectrum* is one where not all wavelengths are present. A neon sign, for instance, has only a few orange ones present. When a source that produces a continuous spectrum is surrounded by a gas, some wavelengths are absorbed. This shows up as a series of dark lines and is called an *absorption spectrum*.

WAVE MOTION

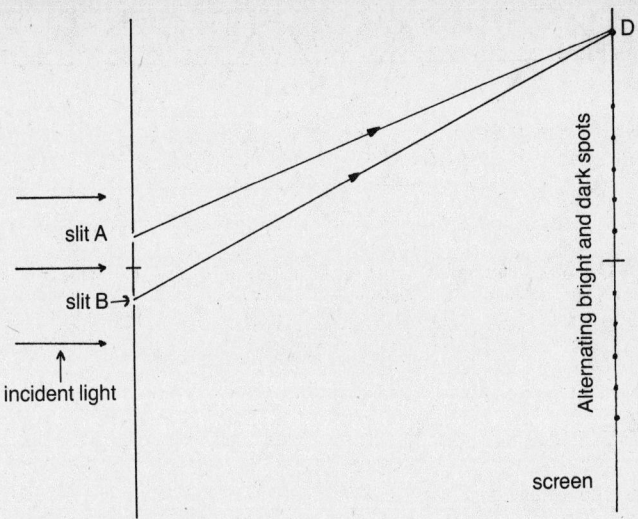

Fig. 6–13 Young's diffraction experiment

Scattering

Scattering occurs when a molecule absorbs light and re-emits it in a different direction. This phenomenon is responsible for the characteristic color of the sky. Short wavelengths (blue) are scattered more effectively by air molecules than long ones (red). It is the scattered light that we see; hence, the sky appears blue. The sun's light, having had some of its short wavelengths scattered, is somewhat biased toward the long waves; the sun appears redder than if no atmosphere were present. At dusk or dawn, the sun's rays travel through the greatest possible amount of atmosphere, making scattering a maximum. Hence, sunrises and sunsets appear even redder.

Young's Experiment

Figure 6-13 shows Young's setup for demonstrating diffraction of light. Light of a particular wavelength is incident on an obstacle with

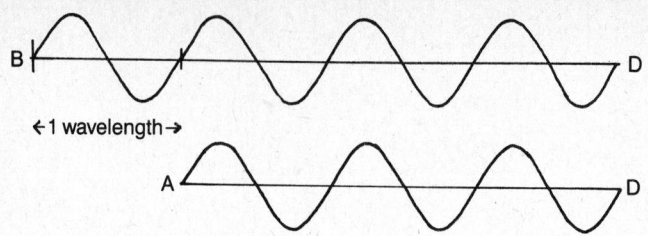

Fig. 6–14 Waves starting out in phase at A and B end up in phase at D if the paths differ by one or any integral number of wavelengths.

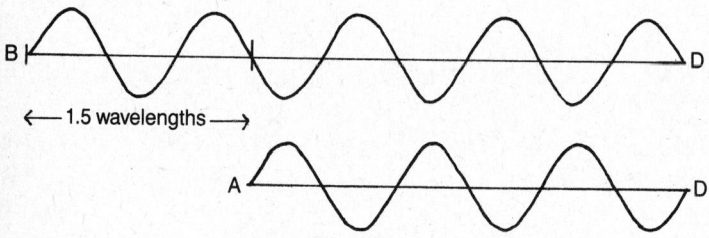

Fig. 6–15 Waves starting out in phase at A and B end up completely out of phase at D if the paths differ by any half-integral number of wavelengths.

two narrow slits. By narrow, we mean that the width of the slit is approximately equal to a wavelength. Beyond this is a screen. Without knowing anything about waves, we would expect only two bright spots on the screen, located opposite the slits. This is not the case, however. The screen shows alternating bright and dark spots, spreading out from a bright spot in the center.

The narrow slits act like point sources; light coming through them beams out in all directions. At an arbitrary point on the screen, D, light rays come from both slits and interfere. Dark spots result from destructive interference, while bright spots result from constructive interference. The type of interference depends on the phase of the waves. We know that the waves start out at A and B in phase. What happens when they arrive at D? That depends on the difference between the number of wavelengths that can fit on the paths AD and BD. When the paths differ by an integral number of wavelengths the waves will end up in phase at D. This is shown in figure 6-14, where the paths differ by a full wavelength. When the paths differ by a half-integral number of wavelengths, the waves will end up completely out of phase at D.

WAVE MOTION

This is seen in figure 6-15, where the paths differ by 1.5 wavelengths It should be evident that as the position of D changes, the path difference changes. As one moves out from the center, the path difference alternates between integral and half-integral wavelength values. Hence, the interference alternates between constructive and destructive, resulting in alternating bright and dark bands. It should be noted that the central bright spot is the brightest; as one goes farther from the center, the bright spots get dimmer.

Polarization

Polarization refers to the orientation of the electric field vector of a light wave. *Polarized* light has its electric field oriented in a particular direction. *Unpolarized* light has components whose electric field vectors are oriented in all directions. A *polarizing filter* transmits light with only a particular polarization.

1) If unpolarized light is incident on a polarizing filter, only that part of the light whose electric field vector is lined up with the filter's direction of polarization will get through.

2) If polarized light is incident on the same filter, it will not get through unless its electric field happens to be lined up with the filter's direction of polarization.

When light is reflected from shiny objects like glass or water, it becomes somewhat polarized. If you wear polarizing sunglasses, which are essentially polarizing filters, the polarized components are stopped. This is because the polarization of the reflected light is unlikely to match that of the glasses.

OPTICS

In this section, we will investigate the prism, lens, and mirror. The prism and lens work by refraction. The speed of light in glass is less than in air. Hence, the air-glass boundary is the site of refraction. The mirror, obviously, works by reflection. The mirror we will study, however, is not the usual flat one, but a curved one.

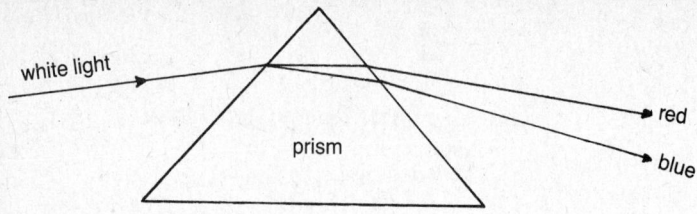

Fig. 6–16 A prism spreads white light into a spectrum of colors because blue wavelengths are bent more sharply than red.

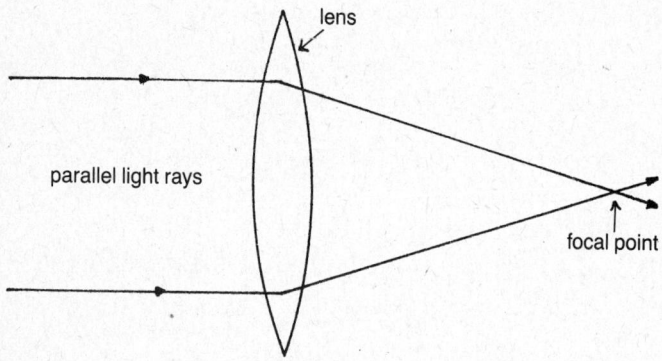

Fig. 6–17 Converging lens

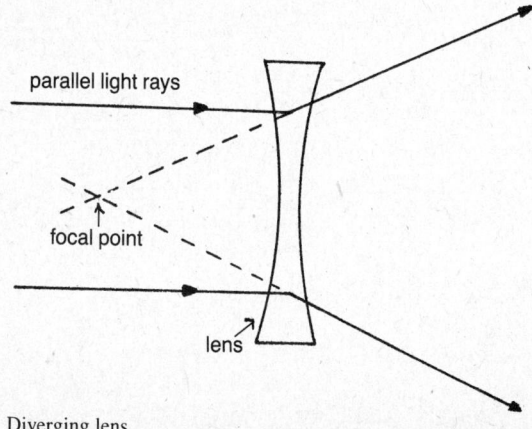

Fig. 6–18 Diverging lens

WAVE MOTION

The Prism

A *prism* is basically a slab of glass as pictured in figure 6-16. When white light goes through a prism, it gets refracted twice because of the two air-glass boundaries. Short wavelengths are refracted more sharply than long ones. Hence, the emerging light is spread out by color into a spectrum.

Lenses and Mirrors

Lenses and mirrors are classified according to how they refract or reflect parallel light rays.

1) A *convex lens,* also called a converging lens (figure 6-17), causes parallel rays to converge. The point where they converge is called the *focal point.* The distance from the focal point to the lens is called the *focal length.* The line passing through the center of the lens is called the *principal axis.*

2) A *concave lens,* also called a diverging lens (figure 6-18), causes parallel rays to diverge. The focal point for a concave lens is obtained by following the diverging rays back (dashed lines) until they cross. Note that these rays do not actually pass through the focal point.

3) A *concave mirror* (figure 6-19) causes parallel rays to converge at the focal point. This mirror is made by grinding out a cavity whose cross section is in the shape of a curve called a *parabola.*

4) A *convex mirror* (figure 6-20) causes parallel rays to diverge. The focal point for a convex mirror is behind the mirror; no light passes through it. These mirrors are found in elevators and stores because their divergent character provides a wide-angle view.

Image Formation

Lenses and mirrors are useful because they can provide images of objects. The formation of these images is shown graphically by a technique called *ray tracing.* Figure 6-21 shows the images of an arrow formed by a convex lens and a concave mirror. Both are called *real images* because light rays pass through the image. The ray tracing technique is slightly different for lenses and mirrors.

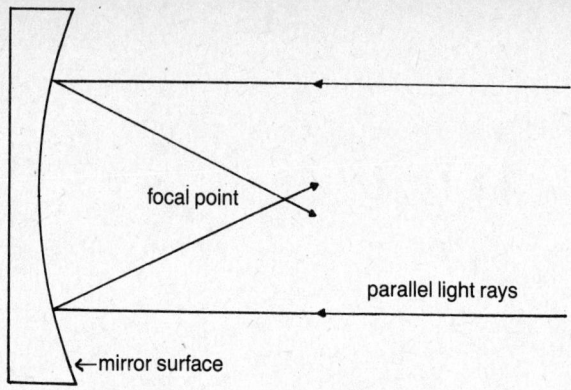

Fig. 6–19 Converging mirror

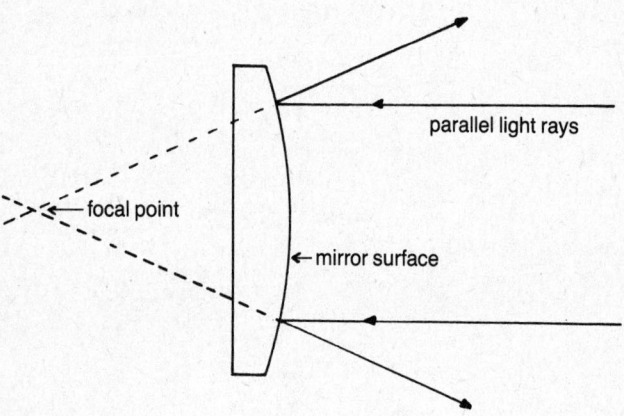

Fig. 6–20 Diverging mirror

1) For lenses, we need three rays:

a) One from the top of the arrow parallel to the principal axis. This gets refracted through the focal point as shown.

WAVE MOTION

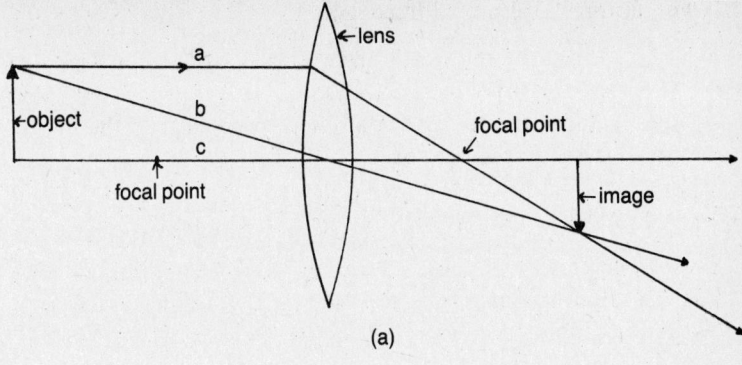

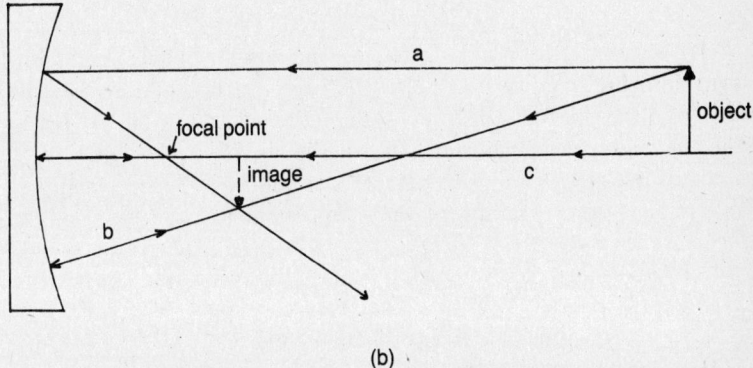

Fig. 6–21 Real image formation by lens (*a*) and mirror (*b*) using ray tracing technique. Rays *a*, *b*, and *c* correspond to those described in the text.

b) One from the top of the arrow through the center of the lens. Any ray through the center of the lens does not get refracted.
c) One from the bottom of the arrow parallel to the principal axis. We have taken the bottom of the arrow to be on the principal axis for convenience, so this ray does not get refracted either.

The two rays from the top of the arrow intersect on the other side of the lens. This is where the image of the top forms. The rest of the arrow is recreated by drawing a line up to the principal axis parallel to the original arrow.

2) For mirrors, we also need three rays:

   a) One ray from the top of the arrow parallel to the axis. This gets reflected through the focal point as shown.
   b) One ray from the top of the arrow perpendicular to the mirror so that it is reflected right back.
   c) One ray from the bottom along the axis. Since this ray is also perpendicular to the mirror, it is reflected right back.

As above, the intersection of the two rays from the top of the arrow locates the top of the image. The rest of the arrow is drawn in the same way as for the lens.

Applications of Lenses and Mirrors

1) *Eyeglasses* correct nearsightedness and farsightedness by refracting light. In figure 6-22, we see a *nearsighted* eye. The lens of the eye brings parallel rays to a focus in front of the retina instead of on it. That is, the eye lens refracts light more than it should. With a diverging lens in front of the eye, figure 6-23, parallel rays will be divergent when they hit the eye, compensating for the overrefraction of the eye. In a *farsighted* eye, figure 6-24, parallel rays can be brought to a focus only at a point behind the eye. With a converging lens in front of the eye, figure 6-25, parallel rays will be convergent when they hit the eye, compensating for the underrefraction of the eye.

2) A *magnifying* glass makes objects appear larger. It produces a *virtual image,* one through which no light passes. Figure 6-26 is a ray diagram showing how a magnifier works. The convex lens is held so that the object is between the lens and the focal point, not beyond the focal point as in figure 6-21a. Note that the rays from the top of the arrow must be extended back to I in order for them to converge. The result is that the viewer thinks the arrow comes from the line labeled image, even though no light actually comes from there.

3) A *telescope* has basically two optical parts: an *objective,* which

WAVE MOTION

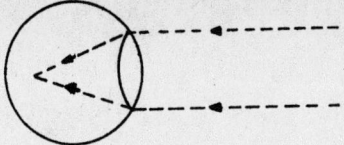

Fig. 6–22 Nearsighted eye

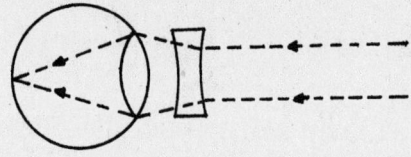

Fig. 6–23 Nearsighted eye with diverging lens

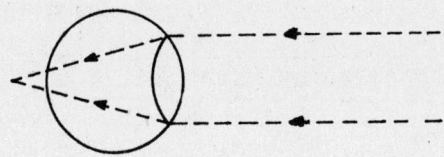

Fig. 6–24 Farsighted eye

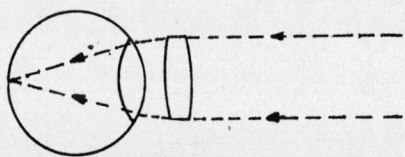

Fig. 6–25 Farsighted eye with converging lens

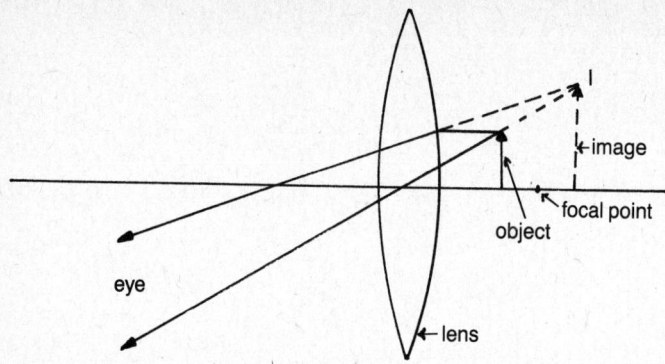

Fig. 6–26 Magnifier

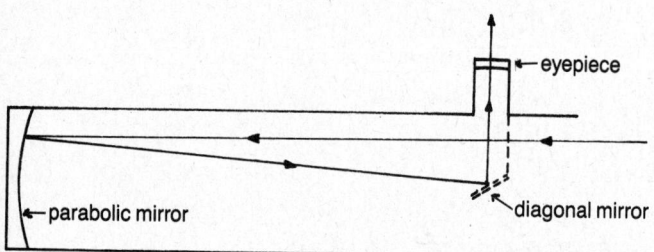

Fig. 6–27 Reflecting telescope

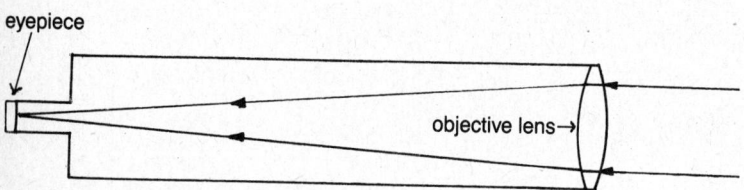

Fig. 6–28 Refracting telescope

WAVE MOTION

forms a real image of a distant object, and an *eyepiece,* which is essentially a magnifying glass.

a) A *reflecting telescope* is shown in figure 6-27. The objective, located at the back of the tube, is a parabolic mirror which focuses parallel rays close to the front of the tube. Just in front of the focal point is a *diagonal mirror* which reflects light into the eyepiece assembly. The eyepiece then magnifies the image.

b) A *refracting telescope* is shown in figure 6-28. The objective is a convex lens in the front of the tube which forms an image at the rear. This image is magnified by the eyepiece.

7

Quantum Theory and Relativity

The physics that we have studied up to now is commonly referred to as *classical physics*. In this chapter, we will look at some problems that classical physics could not solve. This will lead into the study of some revolutionary ideas.

THE DAWN OF THE QUANTUM THEORY

The Question of Radiation

It is a known fact that a heated body emits radiation. The glowing embers of a campfire attest to this. In the middle of the nineteenth century, scientists became especially interested in this radiation.

1) A *blackbody* is a hypothetical object that completely absorbs all wavelengths of radiation which are incident upon it. This absorbtion causes it to heat up and emit radiation whose intensity is different for different wavelengths. The spectrum of this radiation can be measured easily and is shown in figure 7-1. We see that there is one wavelength, λ_o, at which the intensity of the emitted radiation is a maximum. The shape of this curve is characteristic of all blackbodies. The value of λ_o, however, increases as the object cools and decreases as it heats up.

2) In 1880, two physicists, *Rayleigh* and *Jeans,* attempted to explain the blackbody radiation curve. They made the apparently reasonable assumption that energy is a continuous quantity, i.e., it can be absorbed or emitted in arbitrarily small amounts. Their calculations resulted in a curve that fit the actual case for values greater than λ_o. However, as can be seen in figure 7-2, they were way off in the ultraviolet region.

QUANTUM THEORY AND RELATIVITY

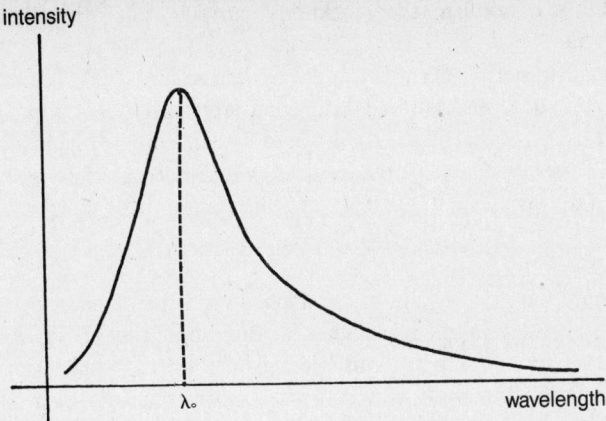

Fig. 7–1 Typical blackbody radiation curve

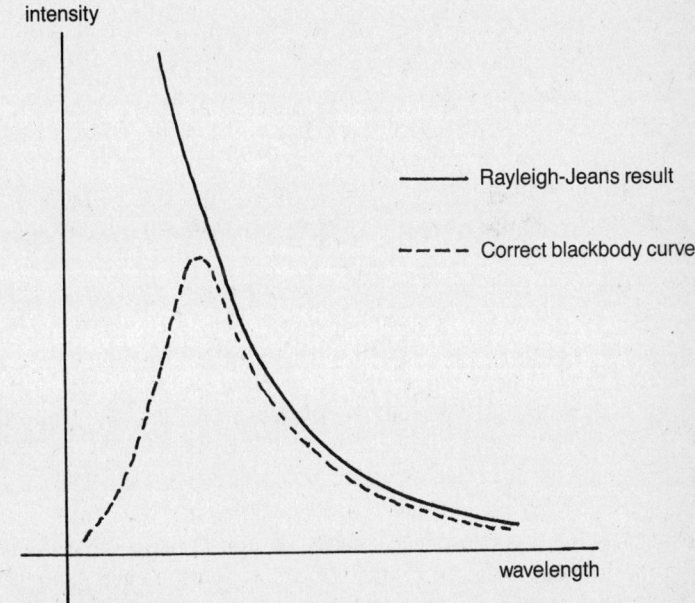

Fig. 7–2 Ultraviolet catastrophe

This failure to explain the blackbody curve is called the *ultraviolet catastrophe*.

3) The physicist *Max Planck* hypothesized in 1900 that radiant energy *cannot* be emitted or absorbed in arbitrarily small amounts. He stated that emission of energy by matter can occur only in discrete bundles called *quanta*. If the frequency of the radiation is ν, then energy is some multiple of a quantity which is proportional to ν:

$$E = nh\nu \quad (n = 1, 2, 3, \ldots)$$

The letter h is a constant called *Planck's constant* and appears frequently in quantum theory. Planck's assumption that energy is a discrete quantity, not a continuous one, enabled him to explain the blackbody radiation curve. It was so radical, however, that even Planck considered it only temporary, for use in the blackbody problem. As we shall see, the idea became quite permanent.

The Photoelectric Effect

1) In 1887, *Heinrich Hertz* shone ultraviolet radiation on a piece of zinc and discovered that electrons were emitted. This phenomenon became known as the *photoelectric effect* and was soon found in other metals using visible light. It was discovered that below a certain frequency, called the *cutoff frequency*, ν_o, even intense light could not produce the effect. Yet at ν_o or above, even a weak light could liberate electrons from metal. In fact, the greater the frequency above ν_o, the greater the kinetic energy of the electrons leaving the metal. This behavior was seemingly inexplicable.

2) In 1905, *Einstein* hypothesized that *all* radiation, not just that from blackbodies, comes in discrete bundles of energy which he called *photons*. Each photon has an energy as given by the Planck relation, $E = h\nu$. How does this apply to the photoelectric effect? In order to break away from the metal, an electron needs some minimum amount of energy. Since energy is $h\nu$, there must be some frequency, ν_o, corresponding to the minimum energy. For frequencies less than ν_o, not enough energy is available to break loose an electron. Now, an intense beam of light is one that has many photons striking the metal per unit time. If none of these photons has the requisite energy, no electrons will be liberated. On the other hand, even a weak beam (few photons

QUANTUM THEORY AND RELATIVITY

per unit time) can liberate electrons if the frequency is greater than ν_o. The excess energy is simply absorbed by the newly liberated electron as kinetic energy.

Einstein's successful explanation of the photoelectric effect established the *dual nature of light:* light exhibits both wavelike and particle-like behavior. For his efforts, Einstein won the Nobel Prize.

Wave-Particle Duality

1) In 1924, *Louis de Broglie* proposed that if light, long held to be a wave, could have properties of a particle, then particles could have properties of a wave. According to de Broglie, this *wave-particle duality* was a fundamental part of nature. He developed a formula for the wavelength of any particle:

$$\lambda = h/p \quad (p = \text{momentum})$$

2) In 1927, two physicists, *C. Davisson* and *L. Germer,* attempted to discover wavelike behavior of particles. They set up an experiment designed to look for diffraction effects in electrons. Recall that in a diffraction experiment, waves are incident on a slit whose width is roughly the same size as the wavelength. Using de Broglie's formula for the wavelength of the electrons in their experiment, they calculated that the slit had to be on the order of the distance between atoms in a piece of metal. They ended up shooting a beam of electrons at a piece of nickel, whose atoms were separated by a distance that was the same as the appropriate slit width. Sure enough, they produced a diffraction pattern. This demonstrated that electrons, which had always been regarded as particles, could also behave like waves. Wave-particle duality became an accepted part of physics.

The Uncertainty Principle

In 1927, *Werner Heisenberg* made yet another startling proposition, his *Uncertainty Principle.* This states that no matter how careful we are in making measurements, there is an inherent limit on the precision with which we can simultaneously measure the position and momentum of a particle. The more accurate we are in measuring position, the less accurate we can be in measuring momentum, and vice versa. Consider the problem of trying to measure the position and momentum of an electron. In order to detect the electron, we must bounce at least one photon off it, just as we must bounce some light off a baseball to see it. Now, if we bounce a photon off an electron, the collision will alter the electron's momentum, resulting in an inaccurate measurement. We can reduce this inaccuracy by using a photon of much lower energy, so that the electron is only minimally disturbed. Since $E = h\nu$, we see that a lower energy means a lower frequency, which means a longer wavelength. But longer wavelengths give poorer resolution. This results in a loss of accuracy in the position measurement. As we can see, trying to improve the momentum measurement fouls up the position measurement. By the same reasoning, trying to improve the position measurement lowers the accuracy of the momentum measurement.

RELATIVITY

Relativity refers to the study of motion as seen by observers in different frames of reference. We have all experienced relative motion. For instance, imagine that you are on a train moving at a speed of 20 m/s and another train passes you at a speed of 25 m/s in the same direction. If you looked out the window at the passing train, you would think it is going at a speed of 5 m/s. However, to an observer on the platform, the other train is moving at 25 m/s.

We might be tempted to say that the observer on the platform has the correct answer because his frame of reference is fixed. A fixed frame is desirable because it has no motion of its own to influence measurements. However, the platform is attached to the earth, which is in motion around the sun. The sun is also moving, as is the solar system. In fact, entire galaxies move with respect to each other. This all boils down to a very important idea: *there are no fixed frames of reference.* The motion of objects has meaning only with respect to a particular frame of reference. In our example above, each answer is correct in its own frame of reference.

QUANTUM THEORY AND RELATIVITY

The Speed of Light

There is one quantity that does *not* depend on the frame of reference: the speed of light in a vaccuum, c. This value is always 3×10^8 m/s. How can this be true in the face of the previous discussion? Let us consider some history.

In 1887, two American physicists, *Michelson* and *Morley,* attempted to measure the relative motion of the earth through the ether. They reasoned that as the earth makes one revolution about the sun, there must be one point where earth and ether move in the same direction. Similarly, there must be another point where earth and ether move in opposite directions. A measurement of the speed of light should be different in these two situations. They expected this difference to show up in a carefully designed experiment. The details of the experiment are not important here. What is important is the result: the effect they expected due to the presumed difference in speeds did not materialize. Despite repeated efforts, they always got the same null results.

It was Einstein who saw the importance of this seeming failure. He believed that Michelson and Morley could not detect a difference in speeds because there simply is no difference. He postulated that the speed of light is the same in all frames of reference. Thus, it doesn't matter how the earth moves with respect to light. To see how extraordinary this idea is, let us reconsider our train example. Suppose that an observer is on a train traveling at a speed close to c. A beam of light passes by the window going in the same direction as the train. Another observer, sitting on the platform, measures the speed of light and gets 3×10^8 m/s, just as we expect. Based on what we saw from the original example, we expect that the observer on the train will get a rather small value for the speed of light since he is going almost as fast as the light and in the same direction. But Einstein's postulate predicts otherwise: the observer on the train gets 3×10^8 m/s, despite the fact that the train is moving. This all seems to defy simple logic.

Special Relativity

Special relativity deals with frames of reference moving with constant velocity. It is here that we can find the answer to our speed of light puzzle.

We know that speed equals distance divided by time:

$$v = d/t$$

We expected the observer on the train to get a rather small value for the ratio d/t. But what if the distance, d, were measured with a meter stick that was shorter than one meter? This would produce a larger d than expected. Similarly, what if t were measured with clocks that were slow? Then t would be smaller than expected. The ratio d/t could conceivably be much larger than we would predict, making a value of 3×10^8 m/s possible. These effects, which we have posed as questions, actually occur.

1) *Time dilation* is the name given to the slowing down of moving clocks. Imagine timing an event with two clocks, one that is stationary and one that is moving. When the dial of the moving clock points to one second, more than one second will have passed according to the stationary clock.

In figure 7-3 we have an example of time dilation. It is a system which consists of two parallel mirrors and a clock. The clock is perfectly synchronized with the mirrors: it ticks whenever a light pulse hits mirror A. The system is set up so that a light pulse leaves A, bounces off B, and returns to A. Hence, the clock ticks once. Figure 7-3a shows what the system looks like when it is at rest with respect to an observer. The path of the light as seen by the observer is indicated by the arrows. In one click's worth, the light has traveled a distance $2d$. Figure 7-3b shows what the system looks like when it is moving with respect to the observer. The light pulse now has a horizontal component of motion, its path again shown by the arrows. This time, in one click's worth, the light has traveled a distance $2D$, clearly greater than $2d$. It takes longer for a tick to occur: hence, the moving clock must tick more slowly than the stationary one.

2) *Length contraction* is the shrinking of a moving object's length. The contraction takes place only along the dimension which is in the direction of motion. If an object is moving with respect to you, it will appear shorter than if it were stationary with respect to you. If you were observing someone in a moving frame of reference making measurements with a meter stick, you would say the meter stick is too short.

It must be noted that time dilation and length contraction are apparent only to an observer looking at a moving system. When the observer is moving along with the system, everything appears normal. Going back to our train example, we note that the observer on the platform thinks that the measurements made on the train are strange.

QUANTUM THEORY AND RELATIVITY

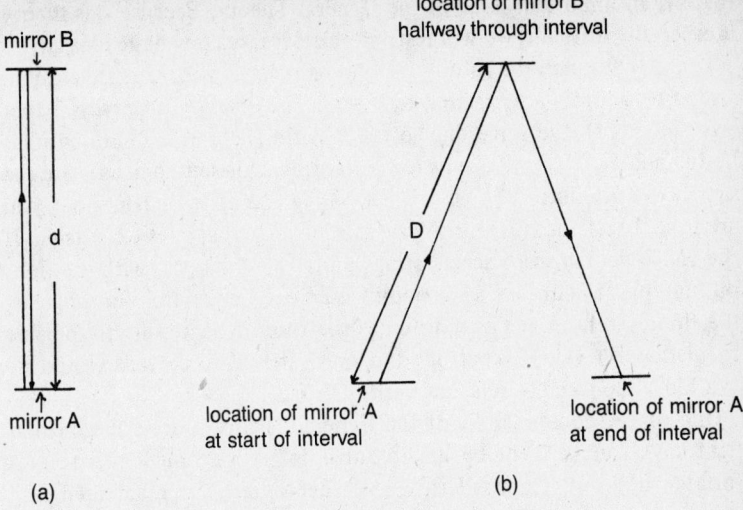

Fig. 7-3 Comparison of stationary clock (*a*) and moving clock (*b*)

The observer on the train doesn't think anything is wrong; it is the measurements made by the observer on the platform that seem awry!

To appreciate what being in different frames means, imagine that you are walking on a crowded street. Suddenly, you are slowed to half your speed. The rest of the world looks as if it is moving too fast. But what if everybody is also slowed to half speed? Then everything looks normal again, even though the speeds are not what they were originally.

For completeness, we should note that the mass of a moving object is greater than the mass of a stationary one. The mass approaches infinity as its speed approaches c. Hence, objects cannot travel faster than the speed of light. It is in connection with mass that we have what is probably the most famous equation in all of physics: $E = mc^2$. This equation symbolizes the equivalence of mass and energy. A small amount of mass can yield a tremendous amount of energy, mainly because the factor c^2 is so large. This is the basis for the extraordinary energy content of nuclear reactions.

General Relativity

General relativity deals with frames of reference that are accelerating. Since acceleration implies force, it is not surprising that the nature of force is an important issue in the General Theory. Because this theory is especially difficult, we will restrict ourselves to a few basic ideas.

The *inertial mass* of a body is that associated with an acceleration, a, due to a force, f, as related by $f = ma$. The *gravitational mass* is that associated with the force due to the gravitational attraction of another body. The *Principle of Equivalence* states that these two masses are the same. We can illustrate this with a simple example. Imagine that you are in a closed room with no way of knowing what goes on outside. If the room were in space accelerating at a rate of 9.8 m/s^2, with the floor pushing up at your feet, your body would exert a reaction force against the floor. From your position inside the room, you would not be able to distinguish this force from the gravitational force that would be exerted if the room were at the earth's surface.

One of the consequences of the General Theory is a prediction that light rays can be bent by gravitational forces. Although this sounds farfetched, an experiment has recently been done which detected such bending of light rays that came near the sun.

IS CLASSICAL PHYSICS WRONG?

We have seen how both quantum theory and relativity have shaken many long-held beliefs about the physical world. Why, then, do we bother with classical physics at all? The answer has to do with the problems being investigated.

Suppose you work in a factory that manufactures 100 lb sandbags for flood control. You put the empty bag on a scale and slowly fill it with sand. The scale indicates an increase from 0 to 100 lb in an apparently continuous fashion. Now, imagine that you are reduced to the size of a bug and the bags are reduced accordingly so they hold, say, 15 grains of sand. As the bag fills, the scale, also reduced to bug size, shows 15 distinct increases in weight.

In both cases, weight is a discrete quantity: the quantum is one grain of sand. To the human, the quantum is insignificant; so what if a bag is three grains over the limit? To the bug, the quantum is very important because it is a significant fraction of weights the bug normally deals with. In this analogy, the grain of sand represents the quantum of energy. The energy of a macroscopic object, such as a car or a baseball, is overwhelmingly large compared to the quantum. In such cases, we can ignore quantum effects. However, when we deal with energies of

QUANTUM THEORY AND RELATIVITY

microscopic systems, such as atoms and molecules, then the quantum is significant; it cannot be ignored. For such systems, classical physics cannot be used.

The effects of relativity in everyday situations are similarly negligible. Relativistic effects change mass, length, and time by a factor of $\sqrt{1 - v^2/c^2}$, called the Lorentz Transformation factor, where v is the speed of the moving object. If the object is at rest, $v = 0$ and the factor is 1, i.e., there are no relativistic effects. Only when v is a sizable fraction of the speed of light will the factor differ appreciably from 1. Since no everyday objects can reach this speed, we don't normally have to worry about relativity; the classical physics principles work quite well.

PART TWO

CHEMISTRY

8
The Atom

Beginning with this chapter, we shall take a closer look at matter, paying particular attention to its microscopic structure and behavior. Ancient Greek philosophers were the first people on record to consider the nature of matter. They were especially interested in whether or not matter could be subdivided endlessly. *Leucippus* and his disciple *Democritus* held that there was ultimately some smallest piece of matter beyond which no subdivision could take place. This was called an *atom*. Unfortunately, most other philosophers, notably *Aristotle*, opposed this idea. They could not believe that there existed a particle which could not be subdivided, so the idea was forgotten. It was not until the nineteenth century that scientists came to accept the idea of an atom, although they knew little about the structure of the atom itself.

THOMPSON'S CONTRIBUTIONS

Cathode Rays

J. J. Thompson, a British physicist, made extensive investigations of a phenomenon known as *cathode rays*. These are produced in an evacuated tube that is constructed with electrodes at each end. When a very high voltage is applied to the electrodes, the negative electrode emits radiation. Thompson identified this as a stream of electrons. In 1897, he measured the ratio of the charge to mass, an important step in ultimately determining the charge itself. (See table 8-1.)

Plum Pudding Model of the Atom

In 1898, Thompson proposed that an atom is a sphere of positively charged material with electrons embedded within, something like raisins in a plum pudding. According to this model, figure 8-1, the total negative charge equals the total positive charge. Thompson's discovery of electrons had shown that atoms contained negative charges. Since it was known that matter was electrically neutral, the negative charge had to be balanced by positive charge. Thompson's model was accepted by scientists because it was in accord with what was known at the time.

TABLE 8-1. ATOMIC PARTICLES

	Electron	Proton	Neutron
charge (coulombs)	-1.6×10^{-19}	1.6×10^{-19}	0
mass (kg)	9.11×10^{-31}	1.67×10^{-27}	1.67×10^{-27}

RUTHERFORD'S CONTRIBUTIONS

Scattering Experiment

In 1911, Rutherford suggested his famous *scattering experiment*. It consisted of aiming a beam of very fast positively charged particles, called alpha particles, at a thin gold foil, figure 8-2. He measured the resultant deviation of the alpha particles from their original path. Based on the Thompson model, there should have been very little deviation. Indeed, most alpha particles went right through the foil without being deviated very much. However, some were unexpectedly deviated through rather large angles. These unusual results led Rutherford to rethink the model of the atom.

Planetary Model of the Atom

Rutherford reasoned that since alpha particles are relatively heavy, only a concentrated positive charge can provide the repulsive electric force needed to deflect them. He suggested that the positive charge of

THE ATOM

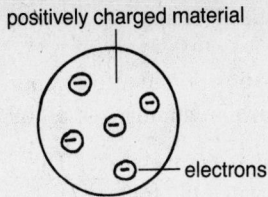

Fig. 8–1 Thompson's model of the atom

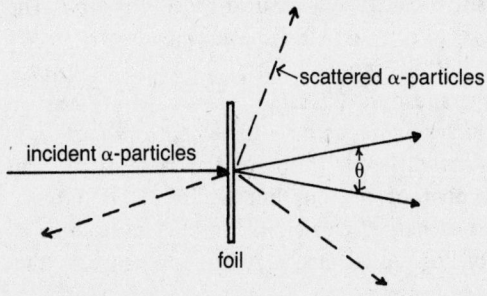

Fig. 8–2 Rutherford's scattering experiment. Rutherford did not expect scattering beyond the range indicated by θ.

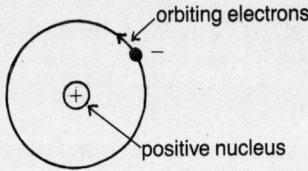

Fig. 8–3 Rutherford's model of the atom

the atom is concentrated in a central region, called the *nucleus,* and the electrons are distributed on the outside, like planets in a solar system, figure 8-3.

Additional Discoveries about Atoms

Other pertinent information about atoms, provided by the work of Rutherford and others, can be summarized as follows:

1) Most of the atom is empty space. This is why so many alpha particles can go through the foil essentially undeviated.
2) All atoms of the same material have the same nuclear charge; atoms of different substances have different nuclear charges. These conclusions come from measurements of the deflection angles of the alpha particles. These angles depend on nuclear charge.
3) The positive charge is a particle called a *proton*. The number of protons in an atom is called the *atomic number*.
4) The nucleus also contains particles called *neutrons*. These have no charge and have the same mass as the protons. All atoms except ordinary hydrogen have some neutrons.
5) *Isotopes* are different forms of the same atom. These have the same number of protons but a different number of neutrons.
6) An *ion* is a charged particle. An atom can become an ion due to a gain or loss of electrons, with no change in the number of protons.

ATOMIC MASS AND ISOTOPES

The *atomic mass* of an atom is defined as the mass of the neutrons plus protons. We ignore the mass of the electrons because they are extremely light in comparison to the protons.

Up to now, we have been measuring mass in kilograms. However, this is too large a unit. A more appropriate unit is the *atomic mass unit* (amu) which is just the mass of a neutron or proton. In amu's, the mass of an atom is a simple number. Nitrogen, for instance, has seven neutrons and seven protons. The atomic mass of a single nitrogen atom is thus 14 amu, or simply 14.

The atomic mass as listed in a table is usually not an integer. This is because when all isotopes of an atomic species are taken into account, their masses have to be averaged according to their relative abundance. For instance, for every four chlorine atoms, three have 17 protons and 18 neutrons. The fourth has 17 protons and 20 neutrons. The average atomic mass is calculated this way:

THE ATOM

 3 atoms like this: 17p + 18n mass = 3 × 35 = 105
 1 atom like this: 17p + 20n mass = 1 × 37 = 37

This averages out to 142/4 = 35.5. Thus, we say that the atomic mass of chlorine is 35.5 even though there is no such thing as half a proton or neutron.

BOHR'S CONTRIBUTIONS

Problems with the Rutherford Model

There are fundamental deficiencies in the Rutherford model.

1) According to electromagnetic theory, a charged particle moving in a circle continuously emits energy in the form of radiation. The electron in an atom should therefore lose energy, circle the nucleus in ever smaller orbits, and eventually collapse into the nucleus.

2) The nature of line spectra, which were believed to be related to atomic structure, could not be explained by the Rutherford model.

The Bohr Model of the Atom

In 1914, Neils Bohr proposed a theory to explain the hydrogen atom. This is the simplest atom, with one proton, one electron, and no neutrons. Although Bohr's calculations are beyond the scope of this book, we can summarize his theory as follows:

1) There are certain orbits in which the electron can circle the nucleus without losing energy. *These are the only orbits allowed.* Each of these orbits corresponds to a particular value of energy, called an *energy level.* The electron is normally in the smallest orbit, which corresponds to the lowest energy. The larger the orbit, the greater the energy.
2) It is possible for some external source to supply the electron with extra energy. If the resulting energy is one of the allowed values, then the electron jumps to a higher energy level. If the resulting energy is not one of the allowed values, the electron remains in the original level.
3) *Line spectra* result when an electron occupying a high energy level loses enough energy to drop back to a lower level. The en-

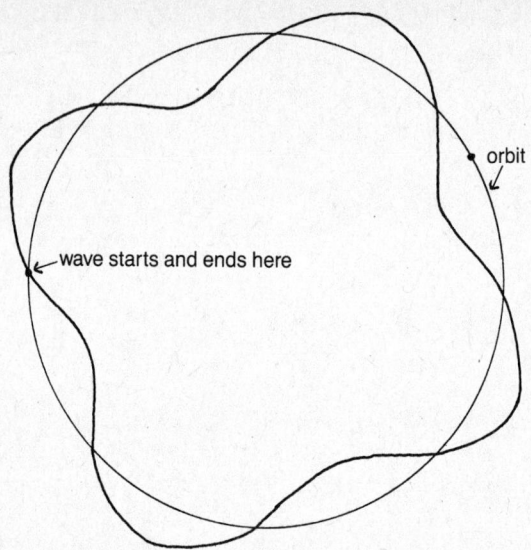

Fig. 8–4 Orbit of exactly four electron wavelengths

ergy difference, E, between the two levels is emitted as electromagnetic radiation. The frequency of the emitted radiation is related to E by the Planck relation, $E = h\nu$.

Interestingly enough, Bohr knew nothing about de Broglie's theory, which came later. Following de Broglie's ideas, we can state that, in order for an orbit to exist, its circumference must be an integral number of electron wavelengths. This is the only way that the end of the wave can be in phase with the beginning when it returns to the starting point (figure 8-4). A nonintegral number of wavelengths would not overlap properly. Destructive interference would take place and the wave (i.e., the orbit) would die out. The orbits calculated by this theory correspond exactly to those calculated by Bohr.

The allowed orbits in the Bohr theory are labeled by the integers 1, 2, 3, etc., which are symbolized by the letter n. The smallest orbit is called $n = 1$, the next is $n = 2$, etc. The letter n is called a quantum number. A *quantum number* is a label which identifies a particular member of the set of allowed values of a discrete quantity, in this case energy. All discrete quantities that we will encounter will be symbolized by quantum numbers.

THE QUANTUM MECHANICAL MODEL OF THE ATOM

The Bohr model cannot explain atoms which have more than one electron. In a multi-electron atom, the many forces of attraction and repulsion are too complicated for the Bohr model. Still, we would like to find out what the allowed energy levels are, just as we did for the hydrogen atom. Also, since there are many electrons, we would like to know how many electrons there are in each level.

The quantum mechanical model rests on two main ideas. The first is the Uncertainty Principle. Because we cannot know the exact position of the electron, we concentrate instead on specifying the probability of finding it in a certain region of space. The second is an equation, called the *Schrödinger equation*, which has in it pertinent information about the atom such as the attractions and repulsions noted above. This equation is solved for a complicated quantity called a *wave function*, denoted by the Greek letter ψ. Buried in this wave function lies very useful information about the atom.

Probabilities

The wave function squared (ψ^2) gives the probability that an electron will be found in a particular region around the nucleus. For every arbitrary region around the nucleus we can actually calculate ψ^2 and find the probability that the electron will be there. Because of the mathematical complexity associated with these probabilities, a geometric representation is used. For each allowed energy state of the electron, a three-dimensional surface can be drawn such that there is a 90 percent chance of finding the electron within the enclosed volume. The length of a vector from the origin to any point on the surface is proportional to the probability of finding the electron in that direction. Two such examples are shown in figure 8-5. In *(a)*, a sphere centered about the origin is shown. The electron has a 90 percent chance of being *somewhere* in that volume (we can't say exactly where). The fact that the surface is a centered sphere means that it is equally likely that the electron will be found in one direction as in another. In *(b)*, there are two spheres along the x-axis. Again, there is a 90 percent chance of finding the electron somewhere inside those spheres. In this case, however,

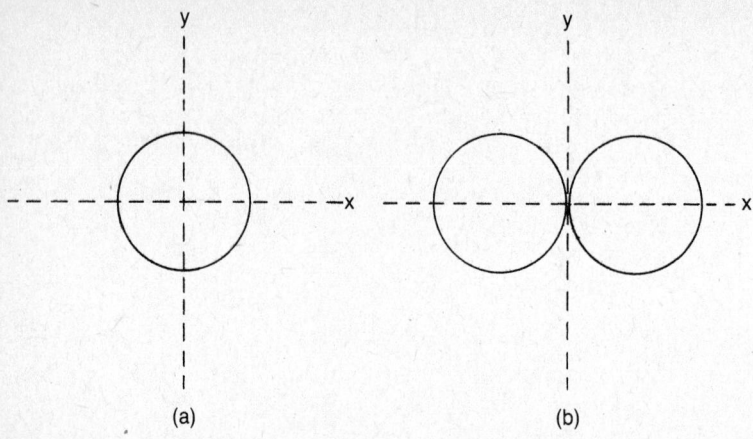

Fig. 8–5 Orbitals. (*a*) S orbital. (b) P orbital.

there is a greater probability of finding the electron along the *x*-axis than in any other direction.

Orbitals

Each allowed energy state of an electron, which we represent by diagrams such as those in figure 8-5, is called an *orbital*. Thus we speak of the size, shape, and orientation of orbitals. In the quantum mechanical theory of the atom, these characteristics are *quantized:* there are only certain allowed sizes, shapes, and orientations. Each characteristic is represented by a quantum number. The size is represented by the number n: the larger the n, the larger the orbital, and the more time the electron spends farther from the nucleus. This is similar to the n in the Bohr model. Another quantum number, ℓ, identifies the *shape* of the orbital. For instance, the shape shown in 8-5a is identified by the value $\ell = 0$. That shown in 8-5b is identified by the value $\ell = 1$. A third quantum number, m_ℓ, identifies the orientation of the orbital. It is important for us to know that the number of orientations depends on the value of ℓ: there are $2\ell + 1$ possibilities.

Orbitals are designated by specifying their size and shape. Thus, each energy level that we are interested in is denoted by a particular combination of the quantum numbers n and ℓ. However, the possible combinations are restricted by some simple rules:

THE ATOM

1) n can be the integers 1, 2, 3, etc.
2) ℓ depends on the choice of n; it can be any integer from 0 up to $n - 1$.

For instance, if $n = 2$, ℓ can be 0 or 1 and the possible n, ℓ combinations are:

$$n = 2, \ell = 0; \quad 2\ell + 1 = 1 \text{ orientation}$$
$$n = 2, \ell = 1; \quad 2\ell + 1 = 3 \text{ orientations}$$

TABLE 8-2. ORBITAL COMBINATIONS FROM $n = 1$ TO $n = 4$

n	ℓ	Possible m_ℓ	Designation
1	0	0	$1s$
2	0	0	$2s$
2	1	-1, 0, +1	$2p$
3	0	0	$3s$
3	1	-1, 0, +1	$3p$
3	2	-2, -1, 0, +1, +2	$3d$
4	0	0	$4s$
4	1	-1, 0, +1	$4p$
4	2	-2, -1, 0, +1, +2	$4d$
4	3	-3, -2, -1, 0, +1, +2, +3	$4f$

There is a more convenient way of denoting these combinations. Each value of ℓ is given a letter:

$$\ell = 0 \text{ called } s$$
$$\ell = 1 \text{ called } p$$
$$\ell = 2 \text{ called } d$$
$$\ell = 3 \text{ called } f$$

We then refer to an orbital by specifying the n-value and the letter: $2s$, $2p$, $3p$, etc. Table 8-2 illustrates the allowed combinations from $n = 1$ to $n = 4$.

Relative Energies

Orbitals increase in energy as the quantity n + ℓ increases. In the event that several n, ℓ combinations have the same sum $n + \ell$, the one with the

TABLE 8-3. ENERGY LEVELS IN ORDER OF INCREASING ENERGY

Designation	$n + l$ value*
$1s$	1 (2)
$2s$	2 (2)
$2p$	3 (6)
$3s$	3 (2)
$3p$	4 (6)
$4s$	4 (2)
$3d$	5 (10)
$4p$	5 (6)
$5s$	5 (2)
$4d$	6 (10)
$5p$	6 (6)
$6s$	6 (2)
$4f$	7 (14)
$5d$	7 (10)
$6p$	7 (6)
$7s$	7 (2)

*Numbers in parentheses represent the maximum number of electrons that can occupy the level.

higher n has the higher energy. Table 8-3 shows the orbitals in order of increasing energy. This order is different from that shown in table 8-2. (The right-hand column shows how many electrons there are in each orbital. This will be explained shortly.)

Electron Spin and the Pauli Exclusion Principle

If we consider the electron to be a spherical charge spinning about its own axis, then there are only two ways it can spin: clockwise or counterclockwise. Thus we have yet another quantum number representing these two possibilities. This is called m_s, and has two values, ½ and -½.

The *Pauli Exclusion Principle* says that no two electrons in an atom have the same identical set of quantum numbers. It can be shown that this principle, coupled with the various quantum number rules, yields a formula for figuring out the number of electrons that can be in a particular orbital:

number of electrons in an orbital = $2(2l + 1)$

We used this formula to get the right-hand column of table 8-3.

THE ATOM

Electronic Configuration

Electronic configuration is a symbolic way of denoting how electrons are normally arranged by energy in an atom. Recall that in the hydrogen atom, the lone electron normally occupies the lowest level. The same idea holds for a multi-electron atom: electrons fill the lowest level first, then the next lowest, etc., until there are no more electrons to put in. For instance, the chlorine atom has 17 electrons. Looking at table 8-3, we see that the 1s, 2s, 2p, and 3s orbitals are completely filled with the first 12 electrons. The next 5 go into the 3p orbital. The arrangement is denoted with its own shorthand system:

$$Cl:\ 1s^2 2s^2 2p^6 3s^2 3p^5$$

The number and letter indicate the orbital; the superscript indicates how many electrons there are in each one. Table 8-4 shows the electronic configuration of various elements that we will have occasion to consider in the next chapter.

TABLE 8-4. ELECTRONIC CONFIGURATION OF SELECTED ELEMENTS

Al: $1s^2 2s^2 2p^6 3s^2 3p^1$
C: $1s^2 2s^2 2p^2$
Cl: $1s^2 2s^2 2p^6 3s^2 3p^5$
H: $1s^1$
Mg: $1s^2 2s^2 2p^6 3s^2$
N: $1s^2 2s^2 2p^3$
Na: $1s^2 2s^2 2p^6 3s^1$
O: $1s^2 2s^2 2p^4$
S: $1s^2 2s^2 2p^6 3s^2 3p^4$

9

Elements and Compounds

In this chapter we will use the quantum mechanical theory of the atom to explain the behavior of chemical substances. To facilitate the discussion, we introduce the concept of shells and subshells. Electrons which have the same n value are said to be in the same *shell*. If those electrons also have the same ℓ value, they are said to be in the same *subshell*. The shell with the highest n is called the *outer shell*.

ELEMENTS

An *element* is a substance whose atoms all have the same atomic number. Each element has a unique set of properties which distinguishes it from all other elements. However, elements can be put into two broad categories: metals and nonmetals. Some important properties of metals and nonmetals are listed in table 9-1.

TABLE 9-1. COMPARISON OF METALS AND NONMETALS

Metals	*Nonmetals*
Characteristic luster	Lacking in luster
Good conductor of heat and electricity	Poor conductor of heat and electricity
Malleable and ductile	Solids are usually brittle
Narrow range of colors	Variety of colors

Periodic Table

The *periodic table*, shown in the front of the book, is a chart of all known elements in order of increasing atomic number. Each entry has

ELEMENTS AND COMPOUNDS

the element symbol along with the atomic number and mass. *Example:* The element chlorine is denoted by its symbol, Cl. Above the symbol is its atomic number, 17; below, its atomic mass, 35.453. For convenience, we have also listed the name of the element, its symbol, and its atomic number in table 9-2. The periodic table has rows, called

TABLE 9-2. ELEMENTS, SYMBOLS, AND ATOMIC NUMBERS

Element	Symbol	#	Element	Symbol	#	Element	Symbol	#
Actinium	Ac	89	Hafnium	Hf	72	Praseodymium	Pr	59
Aluminum	Al	13	Hahnium	Ha	105	Promethium	Pm	61
Americium	Am	95	Helium	He	2	Protactinium	Pa	91
Antimony	Sb	51	Holmium	Ho	67	Radium	Ra	88
Argon	Ar	18	Hydrogen	H	1	Radon	Rn	86
Arsenic	As	33	Indium	In	49	Rhenium	Re	75
Astatine	At	85	Iodine	I	53	Rhodium	Rh	45
Barium	Ba	56	Iridium	Ir	77	Rubidium	Rb	37
Berkelium	Bk	97	Iron	Fe	26	Ruthenium	Ru	44
Beryllium	Be	4	Krypton	Kr	36	Rutherfordium	Rf	104
Bismuth	Bi	83	Lanthanum	La	57	Samarium	Sm	62
Boron	B	5	Lawrencium	Lr	103	Scandium	Sc	21
Bromine	Br	35	Lead	Pb	82	Selenium	Se	34
Cadmium	Cd	48	Lithium	Li	3	Silicon	Si	14
Calcium	Ca	20	Lutetium	Lu	71	Silver	Ag	47
Californium	Cf	98	Magnesium	Mg	12	Sodium	Na	11
Carbon	C	6	Manganese	Mn	25	Strontium	Sr	38
Cerium	Ce	58	Mendelevium	Md	101	Sulfur	S	16
Cesium	Cs	55	Mercury	Hg	80	Tantalum	Ta	73
Chlorine	Cl	17	Molybdenum	Mo	42	Technetium	Tc	43
Chromium	Cr	24	Neodymium	Nd	60	Tellurium	Te	52
Cobalt	Co	27	Neon	Ne	10	Terbium	Tb	65
Copper	Cu	29	Neptunium	Np	93	Thallium	Tl	81
Curium	Cm	96	Nickel	Ni	28	Thorium	Th	90
Dysprosium	Dy	66	Niobium	Nb	41	Thulium	Tm	69
Einsteinium	Es	99	Nitrogen	N	7	Tin	Sn	50
Erbium	Er	68	Nobelium	No	102	Titanium	Ti	22
Europium	Eu	63	Osmium	Os	76	Tungsten	W	74
Fermium	Fm	100	Oxygen	O	8	Uranium	U	92
Fluorine	F	9	Palladium	Pd	46	Vanadium	V	23
Francium	Fr	87	Phosphorus	P	15	Wolfram	W	74
Gadolinium	Gd	64	Platinum	Pt	78	Xenon	Xe	54
Gallium	Ga	31	Plutonium	Pu	94	Ytterbium	Yb	70
Germanium	Ge	32	Polonium	Po	84	Yttrium	Y	39
Gold	Au	79	Potassium	K	19	Zinc	Zn	30
						Zirconium	Zr	40

periods, and columns, called *groups.* The jagged line separates metals, which are to the left, from nonmetals. Despite the fact that atoms get more complex as the atomic number increases, there are periodic recurrences of certain chemical properties. The arrangement of the table is such that elements in a particular group display similar chemical behavior. Here are some noteworthy groups.

1) The *alkali metals* are the elements in group IA, with the exception of hydrogen. They tend to combine very readily with other elements.

2) The *alkaline earths* are found in group IIA. They do not combine with other elements as readily as the alkali metals. However, their strength, hardness, and melting points are higher.

3) The *coinage metals* are in group IB. They are particularly good conductors of electricity and heat. They are also very malleable and ductile.

4) The *halogens* are the elements in group VIIA. Normally, fluorine and chlorine are gases, bromine is a liquid, and iodine and astatine are solids. The halogens tend to combine readily with other elements.

5) The *noble gases* are in group 0. They almost never combine with anything. They are often called *inert* gases.

Importance of the Outer Shell

An element is considered *stable* if it is normally found uncombined with other elements. The only elements that meet this criterion are the inert gases.

In 1916, Lewis and Langmuir speculated that the stability of the inert gases is due to their electronic structure. Consider these three inert gases:

$$\text{Ne: } 1s^2 2s^2 \underline{2p^6}$$
$$\text{Ar: } 1s^2 \overline{2s^2 2p^6} \, 3s^2 \, \underline{3p^6}$$
$$\text{Kr: } 1s^2 2s^2 2p^6 \overline{3s^2 3p^6} \, \underline{4s^2} \, 3d^{10} \, \underline{4p^6}$$

Each of these atoms has eight electrons in its outer shell, which is underlined. A look at the electronic configuration of *all* elements reveals that there are never more than eight electrons in the outer shell. Hence, when an atom has eight electrons in the outer shell, that shell is said to be *filled.* (Exception: The $n = 1$ shell is filled with only two electrons.) According to Lewis and Langmuir, the filled outer shell

ELEMENTS AND COMPOUNDS

is responsible for the stability of the inert gases. All elements whose atoms have unfilled outer shells must exist in combination with other atoms in order to fill them. In view of the above discussion, it is not surprising that the number of electrons in the outer shell determines how elements combine with each other. For the elements used here as illustrative examples, we have listed the number of outer shell electrons in table 9-3. This information comes from the electronic structures listed in table 8-4. Later on in this chapter, we will learn how to use the periodic table to get the same information at a glance.

TABLE 9-3. NUMBER OF OUTER SHELL ELECTRONS OF SELECTED ELEMENTS

Al: 3	H: 1	Na: 1
C: 4	Mg: 2	O: 6
Cl: 7	N: 5	S: 6

COMPOUNDS

A *compound* is a substance which results when the atoms of two or more elements combine and undergo a rearrangement of their electronic structure. The result of this rearrangement is that each atom completes its outer shell. The elements remain combined because they are held together by forces called *bonds*.

Compounds have properties which are different from those of their constituent elements. For instance, chlorine is a poison. Table salt is about 60 percent chlorine, yet table salt is not poisonous.

Compounds are symbolized by *formulas*. A formula is a symbolic notation which indicates how many atoms of each element are needed to form the compound. It consists of a symbol representing the element, followed by a subscript indicating how many of those atoms are needed. For instance, water has two hydrogens and one oxygen. Its formula is H_2O.

Ionic Bonds

In an *ionic bond,* electrons are transferred from one atom to another. The atom that gives up the electrons becomes a positive ion; the atom that takes on the electrons becomes a negative ion. The bond is due to the attraction between positive and negative ions. The result of this process is a filled outer shell for both elements. Consider the following examples:

1) Sodium and Chlorine

We see from table 9-3 that sodium has one electron in the outer shell and chlorine has seven. The sodium gives up its outer electron to the chlorine, as shown in figure 9-1. A bond is formed because the sodium becomes positively charged due to the loss of the electron, while the chlorine becomes negatively charged due to the gain. The loss of one electron leaves the sodium with its filled $n = 2$ shell as the outer one. The addition of this electron to the seven already present in the chlorine completes its outer shell. Since only one atom of each element is needed for this bond, the formula is NaCl.

Fig. 9-1 Typical ionic bond formation

2) Magnesium and Oxygen

Magnesium has two outer shell electrons; oxygen has six. The magnesium gives its two electrons to the oxygen. Except for the fact that two electrons are transferred, the process is identical to the above example. Magnesium becomes positively charged and oxygen becomes negatively charged; an ionic bond thus forms. Since one atom of each is needed, the formula is MgO.

3) Magnesium and Chlorine

The two electrons in the outer shell of magnesium cannot both be given to one chlorine atom. Hence, two chlorine atoms are needed. Each

ELEMENTS AND COMPOUNDS

chlorine becomes negatively charged while the magnesium becomes positively charged. The three ions attract each other to form the bond. The formula in this case is $MgCl_2$.

Covalent Bond

In a covalent bond, one or more pairs of electrons are shared. We will demonstrate this by a simple example: the sharing of a pair of electrons by two hydrogen atoms.

We denote a hydrogen atom by its symbol and a dot to represent the lone outer shell electron:

$$H\cdot$$

The joining of two hydrogen atoms is shown this way:

$$H:H$$

When the two hydrogen atoms are a certain distance apart, the electron of each one is attracted by *both* nuclei. This attraction keeps the atoms together. The distance between the atoms is critical. If the atoms are too far apart, the electrons simply orbit their own nuclei and no bond results. If the atoms are too close, the nuclei repel each other and the atoms get pushed apart.

We can regard the pair of electrons as belonging to both nuclei. Hence, on a shared basis, each outer shell has two electrons. Since this is all that is needed to fill the outer ($n = 1$) shell of hydrogen, both atoms have a complete outer shell. The formula for this combination is H_2.

Three additional examples are presented below. In each case, the number of dots in the picture is the sum of the outer shell electrons of each element. Pairs of electrons between two symbols are the ones that are shared. To verify that the outer shells are filled, you must count each shared pair as belonging to both atoms.

1) HCl

$$H:\overset{..}{\underset{..}{Cl}}:$$

2) CO_2

$$\overset{..}{\underset{..}{O}}::C::\overset{..}{\underset{..}{O}}$$

The sharing of two pairs of electrons is called a *double bond*.

3) N_2

$$:N::N:$$

When three pairs are shared we call it a *triple bond*.

Molecules

A *molecule* is the smallest piece of a covalent compound that retains the properties of that compound. A macroscopic piece of a covalent compound consists of many molecules which are attracted to each other by forces called *intermolecular forces*. These are not the same forces that hold the atoms of the molecule together. A macroscopic piece of an ionic compound, however, consists of alternating positive and negative ions, whose attractive forces hold the material together. There are no molecules in an ionic compound.

1) A *diatomic* molecule is one that contains two atoms. Many gases found in nature are diatomic; H_2, N_2, O_2, F_2, I_2, Cl_2, and Br_2 are examples of diatomic molecules.

2) A *polar* molecule is one that behaves as if one part were positive and one part were negative. Actually, all molecules are electrically neutral. Sometimes, there is an unequal sharing of the electron pairs, with the electrons found more often on one side of the molecule than the other. Hence, one side appears more negative than the other. Figure 9-2 illustrates this, schematically. Water, for instance, is a polar molecule.

3) The *molecular weight* is the weight of all the atoms in a molecule.

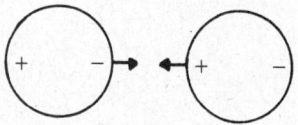

Fig. 9–2 Attraction of two polar molecules. Positive end of one attracted to negative end of the other.

ELEMENTS AND COMPOUNDS

FORMULA PREDICTION

Valence

Valence is a general term which means combining power. *Electrovalence* is the number of electrons given up or taken on by an atom during electron transfer. It is positive if electrons are given up, negative if they are taken on.

1) Elements with 1, 2, or 3 electrons in the outer shell give them up and have electrovalences of 1, 2, and 3, respectively.

2) Elements with 5, 6, or 7 take on enough to make eight in the outer shell. Their electrovalences are 3^-, 2^-, and 1^-, respectively.

Covalence is the number of electron pairs shared by a particular atom in a molecule.

Ionic Formulas

The formula for many ionic compounds can be predicted from the electrovalence. In writing the formula of an ionic compound, the electrovalence of one element is the number of atoms of the other that is needed. The negative sign is ignored in the subscript.

Examples: The electrovalences of some common elements are: $Al:3^+$; $O:2^-$; $Mg:2^+$; $Na:1^+$; $S:2^-$. (We will soon describe a simple method which you can use to determine various electrovalences.) Here are formulas for compounds made from several pairs of these elements:

$$Al \text{ and } O: Al_2O_3$$
$$Na \text{ and } O: Na_2O$$
$$Mg \text{ and } S: MgS$$

In the last example, the formula is not written as Mg_2S_2. The formula for an ionic compound must be written in the lowest terms.

Covalent Formulas

The formula for a covalent compound cannot be predicted in the manner that an ionic compound can. This is because the number of shared pairs often depends on the compound involved. Sulfur and oxygen, for example, can form two compounds as shown:

$$\ddot{\underset{..}{O}}:\underset{..}{\overset{\overset{\displaystyle \ddot{O}}{}}{S}}::\ddot{\underset{..}{O}} \qquad\qquad :\ddot{\underset{..}{O}}:\underset{..}{S}::\ddot{\underset{..}{O}}$$

Clearly, there is no unique formula for a compound made from sulfur and oxygen. For a covalent compound, then, we must know in advance how many of each atom are needed.

POLYATOMIC IONS

A *polyatomic ion* is an ion containing more than one element. The different elements within the ion are held together by covalent bonds; the result, however, is a charged particle capable of forming ionic bonds with other oppositely charged ions. Here are diagrams for two such ions:

Hydroxide: $(OH)^-$

$$\left[:\ddot{\underset{..}{O}}:H\right]^-$$

Ammonium: $(NH_4)^+$

$$\left[\begin{array}{c} H \\ H:\ddot{N}:H \\ \ddot{H} \end{array}\right]^+$$

Note that the number of electrons in the picture for each case is *not* the sum of the outer shell electrons but instead reflects the charge on the ion.

Other important polyatomic ions are: carbonate: $(CO_3)^{2-}$; sulfate: $(SO_4)^{2-}$; chlorate: $(ClO_3)^-$; nitrate: $(NO_3)^-$; phosphate: $(PO_4)^{3-}$.

ELEMENTS AND COMPOUNDS

THE PERIODIC TABLE AND ELECTRONIC STRUCTURE

The features of the periodic table are best explained in terms of the electronic structure of the elements.

1) Elements in the same group have the same number of electrons in the outer shell. From the discussion on bonding, it is clear that the outer shell electrons determine how elements bond. Similarities in the outer shell thus imply similarities in chemical behavior. As an example, sodium combines with chlorine in a one-to-one ratio. Hence, all other elements in group IA combine with chlorine in the same ratio.

2) Groups labeled IA-VIIIA are called *representative elements*. An atom of a representative element is filling either an s or p subshell of the outer shell. The group number tells how many electrons are in the outer shell. Thus, the periodic table tells us at a glance the electrovalence of the representative elements. For instance, Ca is in group IIA, so it has two electrons in the outer shell. Its electrovalence is 2+. Bromine is in group VIIA, so it has seven electrons in the outer shell. Its electrovalence is 1-. The formula for the compound of these two is therefore $CaBr_2$.

3) Groups labeled B are called *transition metals*. An atom of a transition metal is filling an inner d subshell; the outer shell has a filled s subshell. Sometimes the inner electrons are available for bonding, making more than one electrovalence possible. Thus, it is risky to predict the formulas of compounds with them.

4) The separated groups are called *inner transition elements*. These atoms are filling an inner f subshell. We will not concern ourselves with them.

5) For all periods, the n value of the outer shell is the same as the period number.

NAMING SIMPLE COMPOUNDS

Any compound which contains two elements, an element and a polyatomic ion, or two polyatomic ions, can be named by using a simple rule that is modified slightly according to which case applies.

Basic Rule: Name the first element, then name the second using the suffix "ide."

Naming Ionic Compounds

Ionic compounds can be recognized by the fact that one element is a metal (or positive ion) and one is a nonmetal (or negative ion).

1) If both ions are simple elements, then the basic rule is used unmodified. The positive ion is named first.

$MgCl_2$: magnesium chloride
$NaCl$: sodium chloride
Na_2O: sodium oxide

2) If one of the ions is polyatomic its name is used unchanged.

$Mg(CO_3)$: magnesium carbonate
$(NH_4)Cl$: ammonium chloride
$(NH_4)(NO_3)$: ammonium nitrate

3) If the metal can have more than one electrovalence, a roman numeral indicating the charge of the ion is placed between the two names. For instance, the compound made with the Fe^{3+} ion and the O^{2-} ion has the formula Fe_2O_3 and is called iron (III) oxide.

Conversely, we can figure out the electrovalence of the iron from the formula Fe_2O_3. The compound must be neutral. Three oxygens, each with a charge of 2-, contribute a total charge of 6-. The irons must contribute a total charge of 6+ to balance this out. Since there are two irons, each one must have a charge of 3+.

Naming Covalent Compounds

The number of atoms of each element must be indicated by prefixes such as mono (1), di (2), tri (3), tetra (4), and penta (5). Here are some examples: CO_2, carbon dioxide; SO_2, sulfur dioxide; SO_3, sulfur trioxide; P_2O_5, diphosphorous pentoxide.

10
Chemical Equations

When two substances combine chemically, one or more different substances are produced. The original substances are called *reactants* and the new substances are called *products*. The *Law of Conservation of Mass* states that matter cannot be created or destroyed. Thus, each atom of each element present in the reactants must also be present in the products. A *chemical equation* is an expression that shows the products that result from a given set of reactants.

BALANCING EQUATIONS

A *balanced equation* is an equation that shows the correct proportion of reactants and products. If the equation does not show the correct proportions, it is said to be *unbalanced*. Let us consider two equations for the formation of water. It turns out that hydrogen molecules, H_2, and oxygen molecules, O_2, combine to form water, H_2O. Chemically, we write

$$H_2 + O_2 \longrightarrow H_2O$$

As this equation stands, it is unbalanced. It does not obey the Law of Conservation of Mass because there are two oxygen atoms on the left and only one on the right. The equation:

$$\underline{2}H_2 + O_2 \longrightarrow \underline{2}H_2O$$

is the same as the previous one except for the underlined numbers, called *coefficients*. Coefficients denote the relative number of each molecule needed to make the equation consistent with conservation of mass. The 2s that we put in front of each formula multiply the number of atoms in those formulas by 2. Thus, there are four hydrogen atoms and two oxygen atoms on each side. The Law of Conservation of Mass is obeyed, and the equation is balanced. Since all balanced equations obey the same rules, we can make the following general statements:

1) An equation is balanced by picking coefficients needed to make the equation consistent with the Law of Conservation of Mass. This is the *only* way allowed. Often, this is done by trial and error.
2) Subscripts are *never* changed to balance an equation. In our example, we could write:

$$H_2 + O_2 \rightarrow H_2O_2$$

The result is balanced but the substance on the right is not water.
3) The smallest whole number coefficients that balance an equation are the preferred ones. The equation:

$$4H_2 + 2O_2 \rightarrow 4H_2O$$

is balanced, but the coefficients can all be divided by two.
4) Many equations are *reversible*. Either side can be the products or the reactants. A reversible equation is written as follows:

$$A \rightleftarrows B$$

The *forward* direction is from left to right. The *reverse* direction is from right to left.

TYPES OF EQUATIONS

Certain equations can be classified according to the physical processes at work during the reaction.

Combination

In this type, two elements combine to form a compound. Two common examples are

$$4Al + 3O_2 \rightarrow 2Al_2O_3$$
$$2H_2 + O_2 \rightarrow 2H_2O$$

CHEMICAL EQUATIONS

Decomposition

A compound is broken down into its elements. Some simple examples are

$$2PbO \rightarrow 2Pb + O_2$$
$$ZnCl_2 \rightarrow Zn + Cl_2$$

Displacement

One element is displaced from a compound by another. Consider the reaction of zinc and copper (II) sulfate:

$$Zn + Cu(SO_4) \rightarrow Cu + Zn(SO_4)$$

The zinc is originally in its elemental state, and the copper is in the form of a sulfate. The zinc displaces the copper so that the copper is in the elemental state and the zinc is in the form of a sulfate.

Exchange

In this type of reaction, two elements displace each other from their respective compounds. Consider the following:

$$Na_2(SO_4) + BaCl_2 \rightarrow Ba(SO_4) + 2NaCl$$

The barium and the sodium switch places and combine with each other's negative ion.

Combustion

This is the combination of any substance with oxygen. Often we are interested in the combustion of compounds containing hydrogen and carbon. When enough oxygen is present, such reactions always result in the production of CO_2 and H_2O. The combustion of methane illustrates this:

$$CH_4 + 2O_2 \rightarrow CO_2 + 2H_2O$$

MOLES

We have seen from balancing equations that it is sometimes necessary to count atoms and molecules. Of course, they are so small and numerous that counting them one by one would be futile. We must invent a convenient unit that will make counting practical.

A *mole* is the number 6.02×10^{23} (also called *Avogadro's number*) and is the unit we use to count microscopic things such as atoms and molecules. It may seem like a strange choice at first. However, it turns out that 6.02×10^{23} atoms of any element has a weight, in grams, equal to the atomic weight of that element.

Here is an example. The atomic weight of iron is 55.8. Hence, 6.02×10^{23} iron atoms weigh 55.8 grams.

The same idea holds for molecules of a compound. The molecular weight of water is 18. Hence, 6.02×10^{23} water molecules weigh 18 grams.

Since 6.02×10^{23} is clumsy to write, we don't bother with it and use only the word *mole*. Thus we say, for instance:

$$\text{One mole of } H_2O \text{ is 18 grams.}$$

There is nothing special about the *concept* of the mole. You have used it all your life in buying eggs. You buy them by 12s, not one at a time; furthermore, you don't even use the word *twelve* but rather the word *dozen*. What is special is the *value* of the mole. That value makes it easy to find the weight of one mole of anything: just take the atomic (or, if appropriate, the molecular) weight of that substance and tack on the unit grams. If a mole had been defined in any other way, we could not do this.

ENERGY AND CHEMICAL REACTIONS

When a chemical reaction takes place, electrons change positions in order to fill outer shells. Since the potential energy of electrons is a function of their position, changes in position imply changes in potential energy. Thus, in a chemical reaction, there are changes in energy of the reacting system.

CHEMICAL EQUATIONS

Exothermic Reactions

An *exothermic reaction* is one that liberates heat to its surroundings. In this type of reaction, electrons shift position so that their final potential energy is less than their initial. The excess potential energy is liberated as heat.

Endothermic Reactions

An *endothermic reaction* is one that absorbs heat from its surroundings. The potential energy of the electrons is greater after the reaction than before; the extra energy must be absorbed from the outside.

Representing Energy Changes

Let us imagine that two substances, A and B, combine to form two others, C and D. If the reaction were exothermic, we could write the equation as:

$$A + B \longrightarrow C + D + \text{heat}$$

If the reaction were endothermic, we could say:

$$\text{heat} + A + B \longrightarrow C + D$$

Here are two examples with actual measurements:

$$CH_4 + 2O_2 \longrightarrow CO_2 + 2H_2O + 890 \text{ kilojoules}$$

$$CaCO_3 + 178 \text{ kilojoules} \longrightarrow CaO + CO_2$$

The first reaction is the burning of methane, a fuel. It is not surprising that a lot of heat is given off. The second involves the decomposition of calcium carbonate. It takes energy to break this substance down into the given products.

Activation Energy

The *activation energy* is the energy needed to start a reaction going. Even exothermic reactions need to absorb some energy in order to proceed. Consider the reaction:

$$2Na + Cl_2 \longrightarrow 2NaCl + 822 \text{ kilojoules}$$

This reaction liberates a lot of heat. However, some energy must be absorbed before the reaction can take place. This is because the bonds that hold the chlorine molecules together must be broken in order to make Cl^- ions available for bonding to Na^+ ions. Once this happens, the heat liberated by the formation of a small amount of NaCl is more than enough to provide activation energy for the rest of the reaction to proceed.

For some reactions, the activation energy is so small that the energy due to the thermal motion of the molecules is enough to get the reaction going. Also, solutions containing ions need no activation energy; oppositely charged ions attract each other without any outside stimulus.

Catalysts are substances that increase the rate of reaction but do not end up chemically changed. In general, they work by lowering the activation energy needed to get a reaction going. In the human body, many reactions take place in the presence of particular catalysts called *enzymes*.

Stability of Compounds

In general, if a compound requires the absorption of energy for its decomposition, it is stable. Consider hydrogen fluoride:

$$2HF + 538 \text{ kilojoules} \rightarrow H_2 + F_2$$

Unless 538 kilojoules is available to break the HF bonds, they will remain intact and so will the molecule. On the other hand, hydrogen iodide, HI, liberates heat when it decomposes:

$$2HI \rightarrow H_2 + I_2 + 52 \text{ kilojoules}$$

The HI bond tends to break naturally, and the molecule decomposes into the more stable H_2 and I_2 gases.

Rates of Reaction

The *rate* at which a reaction proceeds is a measure of how fast the products are formed. A number of factors influence reaction rates:

CHEMICAL EQUATIONS

1) The higher the *temperature,* the faster the reaction. At high temperatures, more molecules have the necessary activation energy than at low temperatures.
2) If liquids or gases react, increasing the *concentration,* the number of molecules per unit volume, increases the rate of reaction. This just increases the likelihood that reacting particles will meet each other and combine. Stirring has the same effect.
3) For reactions involving solids, increasing the *surface area* increases the rate of reaction. This increases the number of atoms and molecules available for reacting. Grinding a solid into a fine powder can sometimes increase the rate of reaction enormously. Explosions in grain elevators, for example, are due to finely powdered grain combining rapidly with oxygen.

11
Solutions

When two substances are mixed uniformly down to the molecular level, the result is called a *solution*. In this chapter, we shall be mainly, though not entirely, interested in solutions of solids in water, since this is such a common phenomenon. The dissolved solid is called the *solute;* the liquid is called the *solvent*.

MECHANISMS OF SOLUTION

Solution of Ionic Solids

Hydration is the process whereby water molecules, which are polar, surround the individual molecules or ions of the solute. Consider what happens when a piece of NaCl (table salt) is dropped into a container of water. The positive end of the water molecule attracts the negative Cl^- ion; the negative end of the water molecule attracts the positive Na^+ ion. The attraction between water and ion is greater than the attraction of the Na^+ and Cl^- for each other. Hence, the NaCl is pulled apart by the water. Individual Na^+ and Cl^- ions are surrounded by water molecules and held in solution. It should be noted that the symbols Na^+ and Cl^- really stand for things like $Na(H_2O)^+$ and $Cl(H_2O)^-$. However, these formulas are too cumbersome for ordinary use; it is assumed that they are in solution, so we leave off the water part and just use the Na^+ and Cl^-.

Solution of Nonionic Solids

Substances such as alcohol and sugar, whose bonds are not ionic, can also dissolve in water. These molecules are polar; their negative ends are

SOLUTIONS

attracted to the positive ends of the water molecule, and vice versa. Hence, these compounds dissolve. Substances like oil, whose molecules are *not* polar, will not dissolve in water. The polar water molecules have a much greater attraction for themselves than for the nonpolar oil molecules. The attracting water molecules just squeeze the oil into a blob which floats to the top, the oil being less dense than the water.

In general, polar liquids can dissolve polar substances. Nonpolar substances can dissolve only in nonpolar liquids. We say, for short, *like dissolves like*.

DESCRIBING SOLUTIONS

A *saturated* solution contains the theoretical maximum amount of solute at a given temperature. An *unsaturated* solution has less solute than is possible. The *solubility* of a substance is the amount needed to saturate a given quantity of solvent at a given temperature. Under certain conditions, a solution can have more solute than is theoretically possible; it is said to be *supersaturated*. The *molarity* of a solution is the number of moles of solute per liter of solution and is a very important measure of concentration:

$$\text{molarity} = \text{moles of solute/liters of solution}$$

Example: What is the molarity of a solution of NaCl if 58.5 grams is dissolved in 500 ml of solution?

Solution: 58.5 is the molecular weight of NaCl, so 58.5 grams is one mole of NaCl. 500 ml is 0.5 liters. Applying the definition of molarity, we get:

$$\text{molarity} = 1 \text{ mole}/0.5 \text{ liters}$$
$$= 2 \text{ moles/liter}$$

We say this solution is 2 molar, or simply $2M$.

Solutes that break up into ions when they dissolve are said to *dissociate* and are called *electrolytes*. If the dissociation is complete, the electrolyte is called a *strong electrolyte;* if the dissociation is incomplete, it is called a *weak electrolyte*. If no dissociation occurs, the solute is called a *nonelectrolyte*. Electrolytes are important because their mobile ions in solution can conduct an electric current.

FACTORS AFFECTING SOLUBILITY

1) For most solids, an increase in *temperature* of the solution results in an increase in their solubility. For most gases, the reverse is true. That is why carbonated beverages go flat when they warm; the dissolved CO_2 goes out of solution.

2) *Pressure* has an effect mainly on dissolved gases. They become more soluble as the pressure increases. Deep sea divers are affected by this relationship. At great depths, where the pressure is great, gases are very soluble in the blood. If the diver rises to the surface too quickly, the rapid pressure drop forces these gases out of solution too quickly in the form of bubbles in the bloodstream. This condition is known as the "bends."

IONIC EQUATIONS

An *ionic equation* is one which shows how ions in solution react. They are important because many reactions are solutions of compounds which dissociate. Consider the following reaction, which we have already seen and classified as exchange:

$$Na_2(SO_4)(aq) + BaCl_2(aq) \longrightarrow Ba(SO_4)(s) + 2NaCl(aq)$$

We have added some symbols to the equation. The notation (aq) means that the compound is in solution. The notation (s) means that the compound is a solid. When solutions of the two reactants are put together, Na^+, Ba^+, Cl^-, and $(SO_4)^{-2}$ ions are present. Any combination of positive and negative can attract. However, if the compound that results from such an attraction is *soluble,* the compound will not form. If it did, it would just dissolve right away. If the attraction produces a compound that is *insoluble,* the compound will form, since the water molecules are incapable of breaking it apart. It will appear as a solid. In our example, it turns out that the only combination of ions that is insoluble is $Ba(SO_4)$.

We can write the result in terms of the ions present:

$$2Na^+ + (SO_4)^{2-} + Ba^{2+} + 2Cl^- \longrightarrow Ba(SO_4)(s) + 2Na^+ + 2Cl^-$$

Since nothing happened to the Na^+ and Cl^- ions, they are called *spectator ions*. They can be left out of the equation:

SOLUTIONS

$$Ba^{2+} + (SO_4)^{2-} \rightarrow Ba(SO_4)(s)$$

The above equation is called a *net ionic equation*. If all the possible combinations of positive and negative ions are soluble, then all are spectator ions and no reaction takes place. Note that information on the solubility of compounds does not come from the equation. This information must be looked up in a handbook of chemistry.

ACIDS AND BASES

A Closer Look at Water

To a very small extent, pure water undergoes a process called *autoionization*. In this process a hydrogen ion gets attached to one or more water molecules. The result, which is just a hydrated proton, is called a *hydronium ion* and has several possible formulas depending on how many water molecules are involved: H_3O^+, $H_5O_2^+$, $H_7O_3^+$ are all possibilities. For convenience, we refer to all of the above as simply H^+. Then the equation for autoionization of water is

$$H_2O \rightarrow H^+ + OH^-$$

Every liter of pure water has 10^{-7} moles each of H^+ and OH^- ions. Thus for pure water, the concentration of H^+, denoted by $[H^+]$, equals the concentration of OH^-, $[OH^-]$.

Acids

An *acid* is a substance that increases the concentration of H^+ when dissolved in water. All acids contain hydrogen. When an acid is dissolved in water, it dissociates: the hydrogen loses an electron and the remaining proton gets attached to the water molecule to form a hydronium ion. For instance, when HCl dissolves in water, we get the following reaction:

$$HCl + H_2O \rightarrow H_3O^+ + Cl^-$$

As before, we write this without the water:

$$HCl \rightarrow H^+ + Cl^-$$

This does not mean that HCl is made up of H^+ and Cl^- ions. In fact,

HCl is a covalent compound. We have to remember that the above takes place in water and the H^+ stands for hydronium.

Here are some fundamental characteristics of acids:

1) All acids taste sour, liberate hydrogen gas when reacted with certain metals, and cause litmus paper to turn pink. In all acid solutions, $[H^+] > [OH^-]$.

2) *Strong acids* are those that dissociate almost completely, making most of their hydrogen available for increasing the hydronium concentration. Examples are hydrochloric (HCl), sulfuric (H_2SO_4), and nitric (HNO_3).

3) *Weak acids* are those that hardly dissociate at all, producing only a relatively small increase in hydronium concentration. Examples are carbonic (H_2CO_3), acetic ($HC_2H_3O_2$), and boric (H_3BO_3).

Bases

A *base* is a substance that increases the concentration of OH^- ions in solution. Bases usually contain OH^- ions which are liberated as a result of dissociation. Sodium hydroxide is a good example:

$$NaOH \rightarrow Na^+ + OH^-$$

Sometimes a basic solution can be obtained by dissolving a compound which has no OH^- ions. Ammonia gas is an example of this:

$$NH_3 + H_2O \rightarrow NH_4^+ + OH^-$$

Note that OH^- results from this reaction. Some characteristics of bases are:

1) All bases taste bitter and turn litmus paper blue. In all basic solutions, $[OH^-] > [H^+]$.

2) *Strong bases* dissociate completely, liberating a relatively large amount of OH^- ions. Two such bases are potassium hydroxide (KOH) and sodium hydroxide (NaOH).

3) *Weak bases* do not liberate much OH^-. This is often due to the fact that they are only slightly soluble. Even if the little bit that dissolves fully dissociates, as in the case of $Ca(OH)_2$, not much OH^- ion is available.

SOLUTIONS

Salts

A *salt* is a compound that results when an acid and a base are mixed and the solution evaporates until dry. The chemical reaction that takes place during such mixing is called *neutralization*. Consider what happens when a solution of KOH is mixed with a solution of HCl:

$$K^+ + OH^- + H^+ + Cl^- \longrightarrow H_2O + K^+ + Cl^-$$

This solution is neither acidic nor basic because the H^+ and OH^- ions have combined to form water. When the water evaporates, the K^+ and Cl^- ions come together to form the salt, KCl. In solution, they are just spectator ions.

Salts are typically ionic solids and consist of a metal and a nonmetal. They are usually soluble and dissociate completely.

The pH Scale

If the concentration of H^+ ions is expressed in the form 10^{-x} moles per liter, then x is called the *pH of the solution*. This is a convenient way to indicate the acidic or basic character of a solution. We know that for pure water, $[H^+] = 10^{-7}$ moles per liter. Its pH is therefore 7. Solutions having $[H^+] > 10^{-7}$ moles per liter are acidic. Solutions with $[H^+] < 10^{-7}$ moles per liter are basic. Therefore we have the following pH scale:

$pH = -\log [H^+]$

> pH greater than 7: basic
> pH equal to 7: neutral
> pH less than 7: acidic

Accurate measurements of pH can be made with a *pH meter*, an electronic device that indicates pH directly when two electrodes are placed in the solution. Often, an approximate reading is all that is needed. In that case, *indicators* such as dyes or paper tape are used which have characteristic colors in certain pH ranges.

CHEMICAL EQUILIBRIUM

Chemical equilibrium is a situation in which a reaction proceeds in the reverse direction at the same rate it proceeds in the forward direction.

We are particularly interested in how a system in equilibrium responds when the equilibrium is disturbed. *Le Châtelier's Principle* states that when a system in equilibrium is disturbed, the system responds in such a way as to minimize the disturbance. We will investigate this principle by considering the following reaction involving three gases:

$$N_2 + 3H_2 \rightleftarrows 2NH_3 + \text{heat}$$

We will discuss three ways to disturb the equilibrium.

Changing the Amount of a Substance

If we suddenly *increase* the amount of N_2 present, the reaction proceeds in the direction that *decreases* N_2: to the right. This uses up some of the excess N_2, which combines with some H_2 to produce more NH_3. Note that we also end up with less H_2. If we had increased the amount of NH_3 instead of N_2, then the equation would have proceeded to the left, in order to decrease the excess NH_3.

Changing the Pressure

If the pressure on the system is independently increased, the reaction goes to the right. In going to the right, the reaction proceeds from a situation where there are 4 moles of gas to a situation where there are only 2 moles. This can be seen from the coefficients.

To understand why this happens, let us recall the ideal gas law, $PV/T =$ constant. Up to this point, we have said nothing about the "constant" in the equation. We can now be more specific and write it as follows:

$$PV/T = nR$$

Here, n is the number of moles of the gas. R is a number called the *universal gas constant* (the value is not important for this discussion, only the fact that it is constant).

If we do not change the volume or the temperature, then the pressure depends only on the number of moles of gas. The fewer the number of moles of gas, the lower the pressure. Since the equilibrium was disturbed by *raising* the pressure, the reaction must respond by trying to *lower* the pressure, that is, reduce the number of moles. Hence, it must go to the right.

SOLUTIONS

Changing the Temperature

If heat is added to the system, the reaction proceeds to the left. This is because the system responds by proceeding in the direction that reduces the heat: it goes to the left. If heat is removed from the system, it responds by trying to increase the amount of heat: it goes to the right.

$$N_2 + 3H_2 \underset{ENDO}{\overset{EXO}{\rightleftharpoons}} 2NH_3 + heat$$

P: dec., T: inc.

P: inc, T: DEC

$$P \propto n$$
$$T \propto \frac{1}{n}$$

12
Electrochemistry

The relation between electricity and chemistry is a very important one. This is not surprising in view of the role that electrons and ions play in reactions. Some chemical reactions need electricity from an outside source, while others actually produce it. Our main concern in this chapter is a type of reaction called an oxidation-reduction reaction.

OXIDATION-REDUCTION REACTIONS

Oxidation is a process in which a substance *loses* electrons. *Reduction* is a process in which a substance *gains* electrons. An *oxidation-reduction* (redox) reaction is one in which one substance loses electrons at the same time that another substance gains them. Oxidation and reduction occur simultaneously; there is never one without the other.

Half-Reactions

A *half-reaction* is an ionic equation that represents either the oxidation or reduction part of a redox reaction. Consider the reaction of zinc in a solution containing H^+ ions:

$$Zn + 2H^+ \rightarrow H_2 + Zn^{2+}$$

We can see from this that Zn lost two electrons to become a Zn^{2+} ion. This can be represented as follows:

$$Zn \rightarrow Zn^{2+} + 2e^-$$

Similarly, both H^+ ions gained an electron, resulting in H_2:

$$2H^+ + 2e^- \rightarrow H_2$$

When added together, half-reactions yield the original equation.

ELECTROCHEMISTRY

Activity Series

An *activity series* indicates the relative ease of oxidizing or reducing various substances. In table 12-1, the tendency to be oxidized (reduced) decreases (increases) going across the list. This table is useful in predicting whether or not a certain reaction will proceed spontaneously. For instance, when a strip of zinc is placed in a solution containing Cu^{2+} ions, the zinc dissolves and copper metal deposits. Yet, when a strip of copper is placed in a solution containing Zn^{2+} ions, the copper does not dissolve.

TABLE 12-1. ACTIVITY SERIES

K Ca Na Mg Al Zn Fe Ni Sn Pb Cu Ag Au
<----------------------- most readily oxidized ---------------------->
<----------------------- most readily reduced ---------------------->

These observations could have been predicted. What we observe as the dissolution of the zinc strip is, on the microscopic level, the oxidation of zinc:

$$Zn \rightarrow Zn^{2+} + 2e^-$$

When this takes place in the presence of Cu^{2+} ions, the liberated electrons have a choice of going to either species of ion. The activity series predicts that Cu^{2+} ions are more easily reduced. Hence the electrons go to them:

$$Cu^{2+} + 2e^- \rightarrow Cu$$

Copper metal is produced. The whole reaction is the sum of the half-reactions:

$$Zn + Cu^{2+} \rightarrow Zn^{2+} + Cu$$

We can see why copper does not dissolve in a solution containing Zn^{2+} ions. A copper atom would have to lose electrons:

$$Cu \rightarrow Cu^{2+} + 2e^-$$

The electrons would have to prefer Zn^{2+} to Cu^{2+} ions. As we have seen, the activity series predicts the reverse. Hence, the Cu^{2+} ion just picks up the two electrons and becomes a Cu atom again; the copper does not dissolve.

Example: Can Sn dissolve in a solution containing Ni^{2+} ions?
Solution: For $Sn \rightarrow Sn^{2+} + 2e^-$ to take place in the presence of Ni^{2+} ions,

the electrons would have to be picked up by Ni^{2+} rather than Sn^{2+}. The activity series predicts the opposite. Hence, the reaction does not run.

ELECTROLYSIS

Certain reactions which do not occur spontaneously can be made to occur by supplying electricity from an external source. Such a process is called *electrolysis*.

The Electrolytic Cell

An *electrolytic cell* consists of a container filled with an electrolyte, with two electrodes inserted as shown in figure 12-1. The source of ions here is molten NaCl. The two electrodes are attached to a source of DC

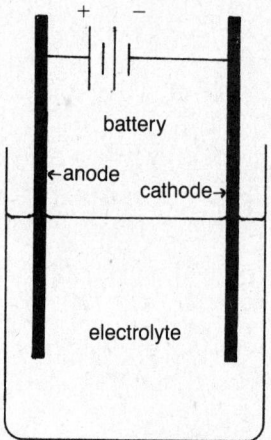

Fig. 12–1 Electrolytic cell

current such as the terminals of a battery. The electrode attached to the positive terminal attracts Cl^- ions and is the site of oxidation:

$$2Cl^- \rightarrow Cl_2 + 2e^-$$

ELECTROCHEMISTRY

Two chloride ions have combined to form a molecule of chlorine gas. The other electrode, which is attached to the negative terminal, attracts Na^+ ions and is the site of reduction:

$$Na^+ + e^- \rightarrow Na$$

A sodium ion has gained an electron to become a sodium atom. By convention, the electrode where oxidation takes place is called the *anode*. The electrode where reduction takes place is called the *cathode*. The two half-reactions added together yield the complete reaction:

$$2Na^+ + 2Cl^- \rightarrow 2Na + Cl_2$$

Electrolysis of molten NaCl is used commercially to prepare sodium. In order to lower the melting point of the electrolyte, $CaCl_2$ is added. Although this introduces Ca^{2+} ions, the activity series predicts that the sodium ions are more readily reduced than the calcium ions. Thus, there is no worry about calcium appearing at the cathode. Interestingly enough, it is not feasible to dissolve the NaCl in water. Water is more easily reduced than sodium. Thus, no sodium would appear at the cathode.

Electroplating

Electroplating is the deposition of a thin layer of metal on top of another metal, using an electrolytic cell. The metal to be plated, such as a spoon, is used as the cathode. The electrolyte contains ions of the metal to be deposited. In this case the spoon, which is the cathode, is dipped in $AgNO_3$. At the spoon, the reaction is

$$Ag^+ + e^- \rightarrow Ag$$

Silver ions from the solution get deposited on the spoon. The anode can be made out of silver if additional silver ions are needed. In that case, the half-reaction is:

$$Ag \rightarrow Ag^+ + e^-$$

The chrome plating of automobile bumpers is accomplished with a solution of $K_2Cr_2O_7$ in sulfuric acid. The bumper is the cathode; the reaction is more complicated than the previous one but is still reduction:

$$K_2Cr_2O_7 + 12H^+ + 12e^- \rightarrow 2Cr + 7H_2O$$

Chromium ions from the $K_2Cr_2O_7$ solution gain electrons and are deposited as Cr atoms on the bumper. The resulting bumper is more resistant to rust and scratches.

VOLTAIC CELLS

We have seen that zinc metal will dissolve in a solution of Cu^{2+} ions. This reaction can be harnessed to produce electricity by means of devices called *voltaic cells,* also called *electrochemical cells.*

Daniell Cells

A *Daniell cell* is a device which can produce electricity by utilizing the electron flow present in oxidation-reduction reactions. Let us consider one made with zinc and copper electrodes.

If zinc metal is placed in a beaker containing $Zn(NO_3)_2$, some of the zinc atoms will start to give up electrons. These electrons can only be taken up by the Zn^{2+} ions in the solution. Thus, there is an equilibrium between the zinc atoms becoming Zn^{2+} ions and vice versa. If copper metal is placed in another beaker containing $Cu(NO_3)_2$, there will be a similar equilibrium between copper atoms and ions. If a wire is placed between the zinc and the copper, as shown in figure 12-2, the electrons are given a choice between going to the Zn^{2+} or Cu^{2+} ions. According to the activity series, the Cu^{2+} ions take on the electrons more readily than the Zn^{2+} ions. Hence, electrons tend to flow from the zinc beaker to the copper beaker. However, as things are pictured in figure 12-2, electrons cannot flow. As soon as a Cu^{2+} ion becomes an atom, the net charge in the copper beaker will be negative, owing to an excess of NO_3^- ions. This inhibits further electron flow. Similarly, when electrons leave the zinc beaker, it becomes positively charged, due to an excess of Zn^{2+} ions. If there were a way to get rid of these excess ions, then current would flow. The *salt bridge,* figure 12-3, solves this problem. This tube is filled with $NaNO_3$ solution and its ends are loosely plugged with glass wool. The end in the copper beaker supplies Na^+ ions to neutralize the excess NO_3^- ions; the end in the zinc beaker supplies NO_3^- ions to neutralize excess Zn^{2+} ions. The neutrality of each beaker is maintained and current flows.

ELECTROCHEMISTRY

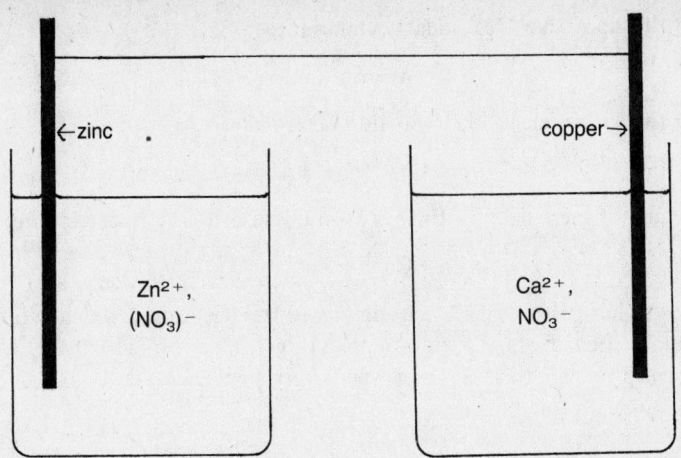

Fig. 12–2 Incomplete electrochemical cell. As soon as an electron flows to the copper beaker, the solution becomes negatively charged, inhibiting further electron flow.

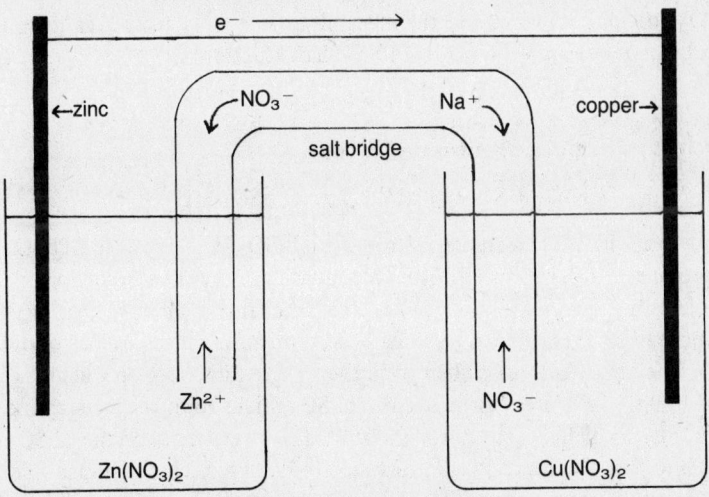

Fig. 12–3 Working electrochemical cell. The $NaNO_3$ in the salt bridge provides Na^+ ions that will combine with excess NO_3^- ions in the copper beaker. It also provides NO_3^- ions that will combine with the excess Zn^{2+} ions in the zinc beaker.

At the zinc electrode, oxidation takes place:

$$Zn \to Zn^{2+} + 2e^-$$

At the copper electrode, reduction takes place:

$$Cu^{2+} + 2e^- \to Cu$$

The sum of these half-reactions is precisely what we have seen before:

$$Zn + Cu^{2+} \to Zn^{2+} + Cu$$

We see how the Daniell cell simply routes electrons, which would naturally flow from the zinc to the copper ions, through a wire. The current produced this way, however, is very small.

Lead Storage Cells

The *lead storage cell* is a rechargeable cell that can produce large amounts of current. It is constructed by interconnecting many cathodes and anodes to form a network with a large surface area. The cathode is PbO_2, the anode is Pb, and the electrolyte is H_2SO_4. There is a loss of electrons at the anode according to the half-reaction:

$$Pb + SO_4^{2-} \to PbSO_4 + 2e^-$$

There is a gain of electrons at the cathode:

$$PbO_2 + SO_4^{2-} + 4H^+ + 2e^- \to PbSO_4 + H_2O$$

Note that the H^+, which comes from the H_2SO_4, is used up and $PbSO_4$ is produced. The loss of H^+ lowers the density of the electrolyte, making density an accurate indicator of the state of charge of the cell.

The lead storage cell can be recharged by operating it as an electrolytic cell. The above reactions then go in the reverse direction, producing H^+ and using up $PbSO_4$. Normally, the $PbSO_4$ sticks to the electrodes and is readily available for electron transfer. However, some of the $PbSO_4$ flakes off as the battery ages, making a complete recharge impossible. Batteries that are completely discharged are particularly susceptible to this flaking and do not last as long as they should.

ELECTROCHEMISTRY

Dry Cells

A *dry cell* is essentially a zinc can filled with an electrolytic paste of MnO_2, NH_4Cl, and carbon. The zinc is the anode and a graphite rod inserted into the paste is the cathode. The half-reactions are

$$Zn \rightarrow Zn^{2+} + 2e^-$$
$$2NH_4 + 2MnO_2 + 2e^- \rightarrow Mn_2O_3 + 2NH_3 + H_2O$$

Note that ammonia is formed. This is in solution and can leak out, causing damage to instruments in which the batteries have been installed. Also, these reactions cannot go appreciably in the reverse direction, so the dry cell cannot be recharged.

13
Organic Chemistry and Biochemistry

Organic chemistry is basically the chemistry of carbon compounds. Biochemistry is the chemistry of life processes. Both subjects are vast; we will treat only the highlights of each.

ORGANIC CHEMISTRY

The remarkable ability of carbon to form bonds is the foundation for organic chemistry. Carbon is a small atom with four electrons in the outer shell. It is exceedingly difficult for carbon either to gain or lose four electrons, so carbon compounds are almost always covalent. What is especially significant is the fact that carbon atoms can bond with each other to form long, stable chains.

Structural Relationships

1) *Structural formulas* are used to represent molecules. These show the relative positions of atoms in a molecule, with the bonds being represented by dashes. For instance, methane has the molecular formula CH_4:

$$\begin{array}{c} H \\ | \\ H-C-H \\ | \\ H \end{array}$$

The number of bonds that each element can form depends on the number of electrons needed to fill the outer shell. For our purposes, we should know that carbon forms four bonds; hydrogen, one bond; oxygen, two; nitrogen, three; and the halogens, one.

2) A *functional group* is simply a particular arrangement of atoms. Functional groups are what give different compounds their different properties. Some examples are

$$\text{Alcohol:} \quad \text{OH}$$
$$\text{Carbonyl:} \quad \text{C}=\text{O}$$
$$\text{Acid:} \quad \underset{\text{OH}}{\text{C}=\text{O}}$$

We will define other groups as we need them.

3) *Isomers* are compounds with different structural arrangements of the same formula. The formula C_2H_6O has two different arrangements, each a different substance:

```
   H   H              H H
   |   |              | |
H–C–O–C–H         H–C–C–O–H
   |   |              | |
   H   H              H H
```

Note that the oxygen is bonded to two carbons in one arrangement, and to one carbon in the other. These are two different bonding arrangements. If we redraw the second formula and interchange the hydrogen on the left with the OH group on the right, we do not create a new isomer; the atoms remain bonded the same way as before.

Hydrocarbons

Hydrocarbons are compounds that are formed from only hydrogen and carbon atoms. There are many types, a number of which we outline below.

1) *Alkanes* have carbon bonded either to hydrogen or to other carbon atoms through single bonds. They have the general molecular formula C_nH_{2n+2}. The first three in this series are shown below:

```
   H           H H          H H H
   |           | |          | | |
H–C–H       H–C–C–H      H–C–C–C–H
   |           | |          | | |
   H           H H          H H H

 methane      ethane        propane
```

Note that each compound has one more CH_2 group than the one preceding it. Larger molecules in this series have a prefix indicating the number of CH_2 groups in the molecule. For instance, *hex*ane has six CH_2 groups:

$$\begin{array}{c} H\ H\ H\ H\ H\ H \\ |\ \ |\ \ |\ \ |\ \ |\ \ | \\ H-C-C-C-C-C-C-H \\ |\ \ |\ \ |\ \ |\ \ |\ \ | \\ H\ H\ H\ H\ H\ H \end{array}$$

All alkanes are *saturated:* they have as many hydrogens attached to the carbons as is possible. Also, they are nonpolar, so they do not dissolve in water.

An *alkyl group* is an alkane with a hydrogen removed. For instance, if ethane had a hydrogen removed, it would look like this:

$$\begin{array}{c} H\ H \\ |\ \ | \\ H-C-C-R \\ |\ \ | \\ H\ H \end{array}$$

R represents something to which the group is bonded.

2) *Alkenes* have at least one carbon-carbon double bond. The simplest one is ethylene and has the formula C_2H_4:

$$\begin{array}{c} H \qquad\quad H \\ \diagdown \quad \diagup \\ C = C \\ \diagup \quad \diagdown \\ H \qquad\quad H \end{array}$$

Other members of this series with *one* double bond have the general formula C_nH_{2n}. The properties of alkenes are similar to those of the alkanes.

3) *Alkynes* have at least one carbon-carbon triple bond. The general formula for this series is C_nH_{2n-2}. The most familiar member of this group is the fuel *acetylene,* C_2H_2:

$$H-C \equiv C-H$$

4) *Aromatic hydrocarbons* are compounds that have a ring of six carbon atoms, pictured below:

Each of the six sides is a bond between carbon atoms; one carbon atom is understood to be located at each apex. The circle indicates that six

ORGANIC CHEMISTRY AND BIOCHEMISTRY

electrons are joined together in a special kind of bond that belongs to the molecule as a whole. This is called a *pi-bond*. The simplest aromatic is called benzene:

$$\underset{H\ \ H}{\overset{H\ \ H}{H-\bigcirc-H}} \quad \text{or just} \quad \bigcirc$$

Other aromatics are naphthalene (mothballs), and toluene (a good carburetor cleaner):

naphthalene   toluene

Compounds like benzene are carcinogenic (cancer-causing).

Compounds with Nitrogen and/or Oxygen

1) *Alcohols* are compounds that result when one or more hydrogens on an alkane are replaced by one or more OH groups. Three common alcohols are shown below:

methanol   ethanol   ethylene glycol

Methanol is a solvent and is poisonous. Ethanol is the alcohol suitable for consumption. Ethylene glycol is the important component in antifreeze.

2) *Aldehydes* are compounds in which at least one hydrogen is bonded to the carbonyl group. An example is the preservative *formaldehyde:*

$$\underset{H}{\overset{H}{\diagdown}}C=O$$

Here, two hydrogens are bonded to the carbonyl group.

3) *Ketones* are similar to aldehydes except that they have only alkyl groups bonded to the carbonyl group. Acetone, a lacquer thinner, has two CH_3 groups bonded to carbonyl:

$$\begin{array}{c} H \quad\quad H \\ | \quad\quad\; | \\ H-C-C-C-H \\ |\;\;\; \| \;\;\;| \\ H \;\; O \;\; H \end{array}$$

4) *Carboxylic acids* are compounds with the acid group:

$$R-C\begin{array}{c}O\\ \| \\ \\ \backslash \\ OH\end{array}$$

Here, R represents any alkyl group. *Lactic acid* is the acid which builds up in the muscles when exercising; this is what causes the feeling of fatigue. *Stearic acid* reacts with sodium hydroxide to form soap. This is an unusual molecule, with polar and nonpolar ends. The nonpolar end can dissolve in nonpolar liquids, like grease. The polar end dissolves in polar liquids, like water. This is why soap can clean.

5) *Esters* are compounds which are the result of reactions between a carboxylic acid and an alcohol. They contain the functional group:

$$\begin{array}{c} O \\ \| \\ C-O \end{array}$$

A well-known ester is *nitroglycerin,* the explosive. Esters are also responsible for the perfumelike aroma of flowers.

6) An *amine* is a saturated carbon compound with an NH_2 functional group bonded to it. An *amide* is the result of a reaction between a carboxylic acid and an amine. Amides have the functional group:

$$\begin{array}{c} O \\ \| \\ C-N \end{array}$$

We will refer to these in the section on biochemistry.

Types of Organic Reactions

1) *Addition reactions* occur when groups of atoms are added to alkenes, alkynes, or aromatics and the carbon-carbon multiple bond breaks. This can be seen in the reaction of ethylene and bromine gas:

ORGANIC CHEMISTRY AND BIOCHEMISTRY

$$\begin{array}{c}H\ H\\|\ \ |\\C=C\\|\ \ |\\H\ H\end{array} + Br_2 \longrightarrow \begin{array}{c}H\ H\\|\ \ |\\H-C-C-H\\|\ \ |\\Br\ Br\end{array}$$

The double bond in ethylene has been broken and two bromines have been added.

2) *Substitution reactions* occur when one or more hydrogens are replaced by other species. The compound ethyl chloride is formed when a chlorine is substituted for a hydrogen in ethane:

$$\begin{array}{c}H\ H\\|\ \ |\\H-C-C-H\\|\ \ |\\H\ H\end{array} \text{ becomes } \begin{array}{c}H\ H\\|\ \ |\\H-C-C-Cl\\|\ \ |\\H\ H\end{array}$$

This compound is sprayed on athletes when they get bruised. It numbs the affected area by chilling it.

3) *Polymerization* is the formation of repetitive, chainlike molecules, called *polymers,* from smaller ones, called *monomers.* We can illustrate this process by considering the polymerization of ethylene. When the carbon-carbon double bond is broken, the result looks like this:

$$\begin{array}{c}H\ H\\|\ \ |\\-C-C-\\|\ \ |\\H\ H\end{array}$$

Electrons on the carbon are now available for bonding. When polymerization takes place, the monomers can bond through these electrons. The result is a chain:

$$\begin{array}{c}H\ H\ H\ H\\|\ \ |\ \ |\ \ |\\-C-C-C-C-\\|\ \ |\ \ |\ \ |\\H\ H\ H\ H\end{array}$$

If fluorines are substituted for the hydrogens on the polyethylene monomer, the result is a monomer of *Teflon:*

$$\begin{array}{c}F\ F\\|\ \ |\\-C-C-\\|\ \ |\\F\ F\end{array}$$

Teflon is used to coat cooking utensils. It has a very high melting point and is very slippery, making it ideal for use in so-called nonstick surfaces.

Other important polymers are *neoprene* (synthetic rubber), *Mylar* (a thin film used to back recording tape), silicones (often used as sealers), and the well-known *nylon*.

4) *Condensation reactions* occur when water is split out from two reacting compounds. This is accomplished when a hydrogen from one compound and an OH group from another combine to form water.

BIOCHEMISTRY

In this section, we will describe the most important types of molecules involved in the life process.

Carbohydrates

A *carbohydrate* is a compound that contains many alcohol groups plus an aldehyde or a ketone.

1) *Monosaccharides* are the simplest carbohydrates. Two examples are the sugars glucose and fructose, illustrated in figure 13-1. In these diagrams, the carbonyl group is circled. In *glucose,* a hydrogen is bonded to the carbonyl group. Hence, it is an aldehyde that is attached to the alcohol groups. In *fructose,* the carbonyl group is attached to groups other than hydrogen. In this case, it is a ketone that is attached to the alcohol groups.

2) *Disaccharides* consist of two monosaccharides. *Sucrose,* for example, is a sugar which is a combination of the monosaccharides glucose and fructose.

3) The *polysaccharides* starch, glycogen, and cellulose are polymers of glucose. *Starch* is a reserve food supply that plants store in roots, stems, and leaves. *Glycogen* is stored in the liver by animals. It becomes a supply of glucose when the glucose level in the blood drops below a certain level, such as during marathon running. *Cellulose* makes up the fibrous parts of plants. Its glucose monomers are held together in such a way that animals cannot digest it.

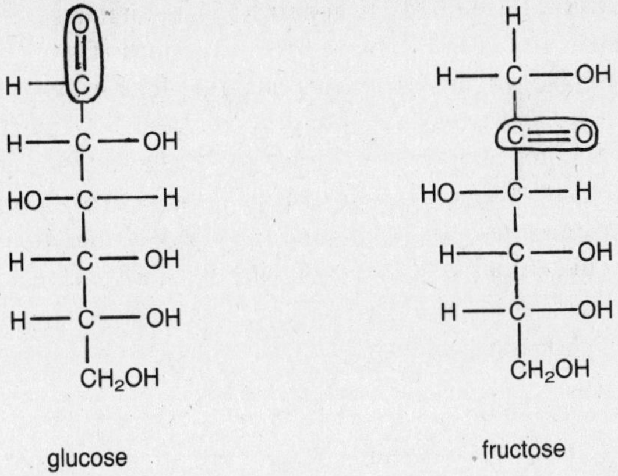

ORGANIC CHEMISTRY AND BIOCHEMISTRY

glucose

fructose

Fig. 13-1 Glucose and fructose

Fats

Fats are esters which come from the reaction of carboxylic acids and the alcohol glycerol. (This particular alcohol has the structure of propane, except that OH groups have replaced three hydrogens.) If the acid molecule is small, the fat is liquid at room temperature and usually is unsaturated. If the acid molecule is large, the fat is solid at room temperature and saturated. The name *oil* is often given to unsaturated fats.

Vitamins

Vitamins are organic substances that are parts of enzymes. They are necessary for growth and maintenance of body functions. In general, vitamins cannot be manufactured by the body, so they must be obtained from outside sources, usually foods. (Exception: when the body is exposed to sunshine, it can make vitamin D.)

Vitamins A, D, E, and K have relatively nonpolar molecular structures.

This means that they can dissolve only in nonpolar substances. Since the fatty tissue of the body is nonpolar, these vitamins can dissolve in it and remain stored in this tissue for long periods of time. Daily intake of a fat-soluble vitamin is not necessary as long as some remains in storage.

The B vitamins and vitamin C have relatively polar molecular structures; therefore, they can dissolve in water. Excesses of these vitamins cannot be stored because they can be passed with the urine. It is best to get adequate supplies of these vitamins on a daily basis.

Proteins

Proteins are made up of amino acids, of which there are about 20 different kinds. In an *amino acid,* an amine group is attached to the carbon atom that is nearest the acid functional group. An example is alanine:

$$\begin{array}{c} \text{H} \quad \text{H} \quad \text{O} \\ | \quad\; | \quad\; \| \\ \text{H}-\text{C}-\text{C}-\text{C}-\text{OH} \\ | \quad\; | \\ \text{H} \quad \text{NH}_2 \end{array}$$

Note that the amine group is on the carbon nearest the acid group. Amino acids can react with each other: the acid group of one reacts with the amine group of the other. The resulting structure ends up with an amine group on one end and an acid group on the other end. Thus, many amino acids can link together. The bonds that accomplish this are called *peptide linkages.*

A *protein* is a chain of at least fifty amino acids. It is the major component of living cells. Because there are so many amino acids in a protein chain, an enormous variety of arrangements is possible. Each variety has its own unique biological properties. One consequence of this is that protein found in one species is usually different from protein found in another. Indeed, members of the same species often have proteins that are slightly different from each other. This is why blood types have to be matched for a transfusion; it is also why rejection can take place during transplants.

ORGANIC CHEMISTRY AND BIOCHEMISTRY

Nucleic Acids

A *nucleotide* is a molecule which consists of a phosphate group, a sugar containing five carbons, and one of four possible nitrogen-containing bases. A *nucleic acid* is simply a polymer whose monomer is a nucleotide. *Ribonucleic acid* (RNA) has a sugar with the formula $C_5H_{10}O_5$. *Deoxyribonucleic acid* (DNA) has one less oxygen in the sugar. In figure 13-2, we see a schematic diagram of DNA. There are two alternating chains of phosphate and sugar groups. Connected to each sugar group is one of the four possible nitrogen bases: adenine (A), guanine (G), cytosine (C), and thymine (T). Pairs of these bases link the two chains together, with AT and CG always coupled. RNA, as will be shown, is assembled from A, G, C, and another base called uracil (U).

The schematic figure shows only a flat picture. In reality, each part of the DNA chain is a helix. The two parts spiral around each other in a form called a *double helix*.

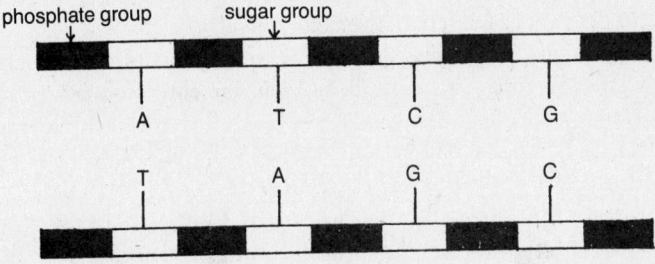

Fig. 13–2 Schematic representation of DNA

The Genetic Code

The *genetic code* is the sequence of pairs of bases in the DNA molecule. As we shall see, it is called a code because the sequence is essentially a set of instructions that cells use to assemble amino acids into particular proteins.

It is important to understand the basics of cell structure. A *cell* (figure 13-3) has three main regions: the *nucleus,* in the center; the *cytoplasm,* surrounding the nucleus; and the *cell membrane,* a wall

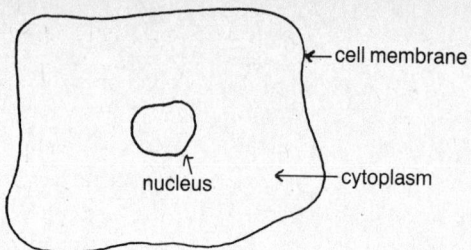

Fig. 13-3 Schematic diagram of a cell

which encloses the entire mass. We are particularly interested in the chemistry involved in the passing of inherited traits from one generation to the next.

The nucleus contains tiny structures called *genes.* These are made of DNA and carry the inherited traits of the organism. Groups of genes are located on larger structures called *chromosomes.* In the human body, each cell has 46 chromosomes which carry thousands of inherited traits such as hair color, eye color, etc.

Cells reproduce by dividing in half. There are two types of nuclear division, the essential difference being how the chromosomes are distributed among the daughter cells:

1) In *mitosis,* each daughter cell receives a set of chromosomes

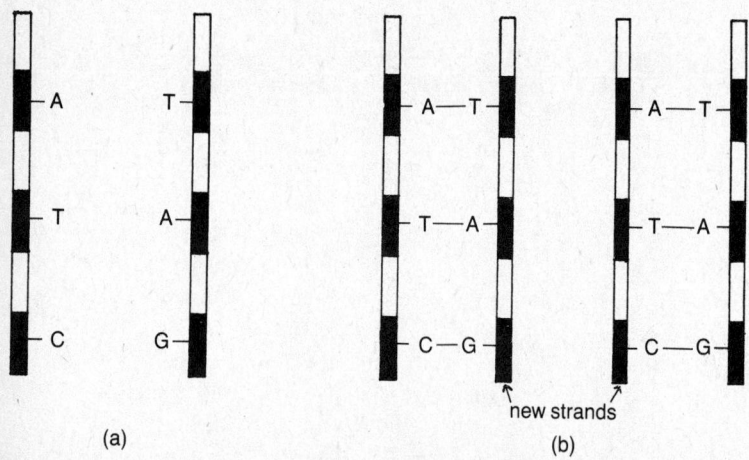

Fig. 13-4 In (*a*), the two strands come apart. Nucleotides floating in the nucleus assemble opposite them, their bases lining up according to the rule for pairing: *AT* or *CG*. Hence, two identical DNA molecules (*b*) are produced.

ORGANIC CHEMISTRY AND BIOCHEMISTRY

exactly the same as the parent. This is shown schematically in figure 13-4.

2) In *meiosis*, each daughter cell gets half the number of chromosomes of the parent. This is the method by which the male sperm and the female egg cells are produced. When the sperm and the egg unite, the correct number of chromosomes is restored. Note that the offspring has chromosomes, and therefore traits, from each parent.

When a cell divides, the coded message in DNA is passed on to the daughter cells in a simple manner. The DNA helix first comes apart like a zipper (figure 13-4a). Each strand rebuilds its missing partner from the nucleotides with various bases that are floating around in the nucleus. Each base on the unzipped strand can pair up with only one type

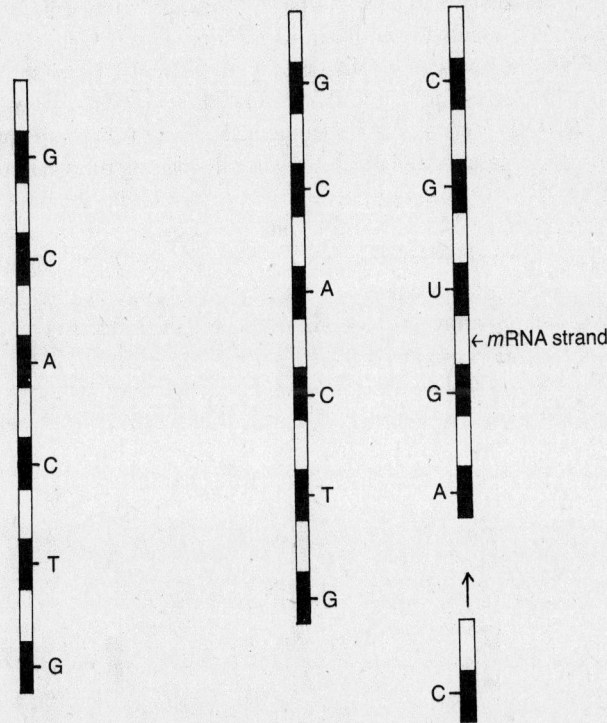

Fig. 13–5 Strand from DNA

Fig. 13–6 Nucleotides at right assemble opposite DNA strand to form *m*RNA.

of unattached base according to the rules of pairing: A with T, C with G (figure 13-4*b*). Hence, the two new DNA chains must be identical to the original.

By means of the genetic code, DNA directs the synthesis of proteins, that is, the assembly of amino acids in the proper sequence for the particular protein being made. The first step in this process is the production of a substance called *messenger RNA,* or mRNA for short. In producing mRNA, the DNA molecule partially unzips itself (figure 13-5). Then the various bases of the nucleotides in the nucleus line up opposite the DNA strand, forming a new strand (figure 13-6). Note that this new strand is *not* attached to the DNA strand. Also, it contains U's, not T's; the U's line up opposite A's. The new strand is the mRNA; when it is completed, it moves into the cytoplasm, carrying the message. Each group of three nucleotide bases codes for a specific amino acid.

In the cytoplasm, the mRNA attaches to a structure called a ribosome, where it meets up with another form of RNA called *transfer RNA* (tRNA). Strands of tRNA, shown in figure 13-7, are short and have a variety of possible combinations of three unpaired bases. Each kind of tRNA attracts a specific amino acid. A tRNA lines up opposite a group of three bases of the mRNA so that the pairing rules are obeyed. Since each tRNA brings with it a specific amino acid, a unique combination of amino acids results from the sequence on the mRNA (figure 13-8). After the amino acid has been deposited in the right place on the growing protein chain, the tRNA breaks away (figure 13-9) so that it can pick up more amino acids and repeat the process.

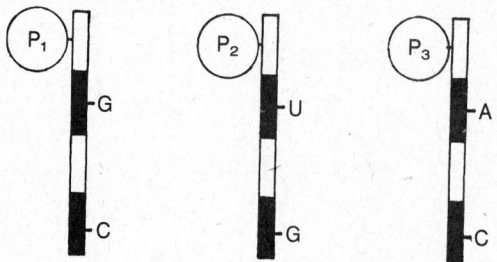

Fig. 13–7 Schematic figure showing three different *t*RNA molecules. Each combination is attracted to a particular amino acid, denoted by P_1, P_2, and P_3.

ORGANIC CHEMISTRY AND BIOCHEMISTRY

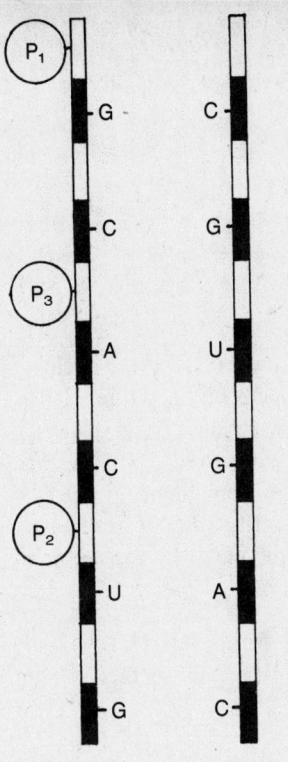

Fig. 13–8 Some *t*RNAs line up opposite *m*RNA, forming a chain of three amino acids, $P_1 P_3 P_2$.

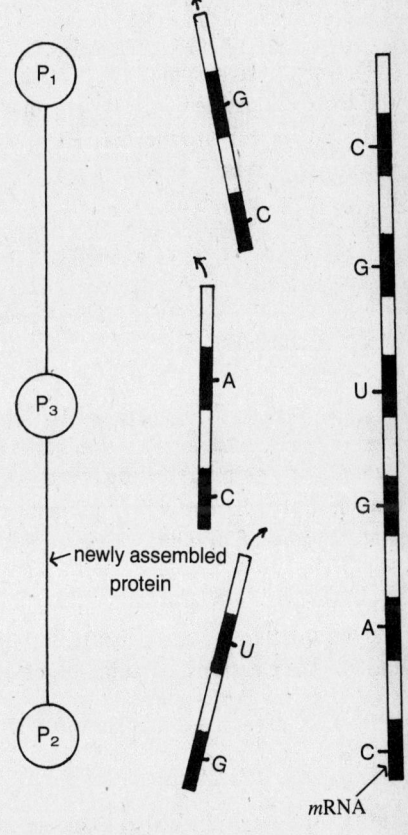

Fig. 13–9 The *t*RNAs break up.

14
The Nucleus

In this chapter, we will look at changes that take place in the nuclei of atoms. These changes are independent of electronic structure. Hence, we will ignore the electrons of the elements we are discussing.

The names and symbols for nuclei are slightly more complex than those for their elements. For a nucleus, the element name must be followed by the atomic mass. This is to identify which isotope is involved. The *symbol* must include both atomic mass and atomic number. Some examples are:

$$_{90}Th^{234} = \text{thorium-234}$$
$$_{83}Bi^{210} = \text{bismuth-210}$$
$$_{92}U^{234} = \text{uranium-234}$$

RADIOACTIVITY

Radioactivity refers to the energy that is spontaneously emitted from a nucleus. Its discovery was accidental; in 1896, *Henri Becquerel* found that a piece of uranium fogged a sealed photographic plate. He also found that various *compounds* of uranium produced the same effect. Now, the electron structure of uranium in a compound differs from that of uranium as an element. Becquerel concluded that the nucleus, and not the electrons, must be responsible for radioactivity.

Types of Radiation

Rutherford determined that there are three types of radiation.

1) *Alpha particles* are the nuclei of helium atoms. Since a helium

THE NUCLEUS

nucleus has 2 protons and 2 neutrons, alpha particles have a charge of +2 and a mass of 4 amu. We can denote an alpha particle by $_2He^4$.

2) *Beta particles* are just electrons. Hence, their charge is -1 and their mass can be considered 0. We can denote a beta particle by $_{-1}e^0$. Of course, there are no electrons in the nucleus. The emission of electrons from the nucleus results from an unusual process in which a neutron becomes a proton and an electron. The proton stays in the nucleus; the electron is emitted.

3) *Gamma radiation* is high-energy electromagnetic radiation. The symbol for gamma radiation is simply γ.

When radiation impinges on substance, it often knocks electrons out of atoms, producing ions. This is what makes radiation so dangerous to living tissue. The well-being of cells depends on an intricate chain of biochemical events. When this chain is upset by the production of strange substances, serious harm can result.

Half-Life

The *half-life* of a radioactive substance is the time required for the mass to decrease to one-half its original value. For instance, if you have 100 grams of a substance whose half-life is 8 days, then after 8 days you have only 50 grams left. After 16 days, you have 25 grams left, etc.

Half-life calculations are used to determine the age of very old objects, such as fossils. Carbon, for instance, has a radioactive isotope with a mass of 14 instead of the usual 12. Living tissue has a constant ratio of carbon-14 to carbon-12. When the tissue dies, the carbon-14 decays, reducing the ratio. Measurement of this ratio, along with a knowledge of the carbon-14 half-life (about 5730 years), allows scientists to figure out age.

Radiation Detectors

1) The *Geiger counter* is basically a tube filled with argon atoms. This is shown in figure 14-1. A thin window at one end admits radiation. The rod in the center is an anode; the outer casing of the tube is a cathode. These are connected to a high voltage source. When the radi-

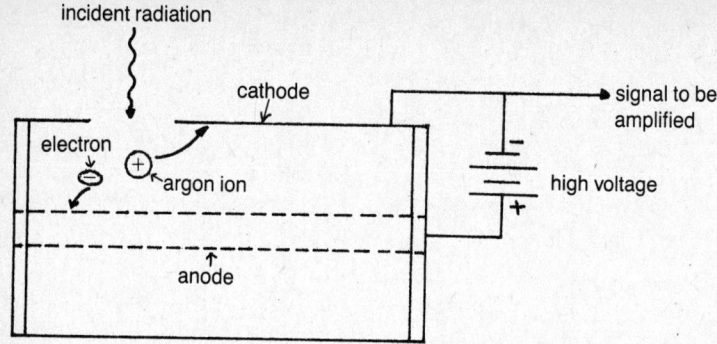

Fig. 14–1 Geiger counter tube

ation enters the tube it knocks out some of the electrons from the argon atoms, producing positive argon ions. The ions are attracted to the cathode, while the electrons are attracted to the anode. The resulting electrical current, which is very small, is amplified and converted to a usable reading.

2) The *cloud chamber* is essentially a container filled with air that has been supersaturated with water vapor. (This means that the air has more water vapor than we expect it to have at that temperature.) When the radiation produces ions, the excess water vapor condenses around them, leaving a visible vapor trail.

3) The *scintillation counter* is a device which depends on the fact that certain materials, called *phosphors,* emit light when radiation is incident upon them. The light is sensed by photoelectric cells which produce electrical signals. These signals are amplified and sent to an appropriate recording device.

Nuclear Equations

The emission of alpha and beta particles from a nucleus results in the formation of different elements. This process is called *radioactive decay* and is usually represented by a *nuclear equation.* This is similar to a chemical equation in that certain quantities, in this case mass and charge, must be the same on both sides of the equation.

1) *Alpha decay* is symbolized by writing the radioactive nucleus on

THE NUCLEUS

the left and the alpha particle and new nucleus on the right. For instance, radium-226 is an alpha emitter. Upon emission, it loses 2 units of charge and 4 units of mass. Since radium has 88 protons, the new substance must have 86 protons. The element with atomic number 86 is radon. Thus, the new nucleus must be a radon nucleus. The equation is:

$$_{88}Ra^{226} \rightarrow {}_2He^4 + {}_{86}Rn^{222}$$

2) *Beta decay* is symbolized in a similar fashion. For instance, cobalt-60 is a beta emitter. Since a charge of -1 is emitted, the new nucleus must have a charge one greater than cobalt. Cobalt has 27 protons, so the new nucleus must have 28. This is a nickel nucleus. The equation describing this is:

$$_{27}Co^{60} \rightarrow {}_{-1}e^0 + {}_{28}Ni^{60}$$

It should be noted that most nuclei that emit these radiations also emit gamma radiation. Since neither the charge nor the mass of the nucleus changes due to gamma emission, the gamma symbol is often left out of the equation.

NUCLEAR ENERGY

The force which holds the nucleus together is called a *short-range force:* it is powerful when nucleons are close together and zero when they are separated. When it acts, the nuclear force is greater than the repulsive electrical forces. If it weren't, the nucleus would just fly apart. The energy associated with the nuclear force, and how we harness it, is the subject of this section.

Meson Theory of Nuclear Force

According to the *meson theory*, protons and neutrons are made up of a central core surrounded by particles called *mesons*. Neutrons and protons have the same core; they differ in the surrounding mesons. Two protons attract each other as a result of exchanging neutral mesons. So do two neutrons. However, a neutron and a proton attract each other by exchanging charged mesons.

Binding Energy

The *binding energy* of the nucleus is the energy needed to break it apart. The mass of the nucleus is less than the sum of the masses of all the separate nucleons. This mass difference, when converted to energy by the relation $E = mc^2$, is the source of the binding energy.

Mathematically, the binding energy is a negative quantity. The negative sign simply means that outside energy must be supplied to pull the nucleons apart. If the sign were positive, it would signify that outside energy is required to hold the nucleons together. In that case, they would not stay together of their own accord; that is, we would have no nucleus. For simplicity, we will call the binding energy between any two nucleons $-B$. Here are two cases:

∞ binding energy = $-B$
∞∞ binding energy = $-3B$ ($-B$ from each pair)

Fusion

Fusion is the formation of a large nucleus from smaller ones. Fusion can produce energy. There is an isotope of hydrogen called *deuterium*, which has one proton and one neutron in the nucleus. When two deuterium nuclei come together, the four particles form a three-particle nucleus and a free neutron. This is shown in figure 14-2. Note that we started with an energy of $-2B$; by conservation of energy, we must end up with an energy of $-2B$. Since the binding energy of the three-particle nucleus is $-3B$, an energy of $+B$ must be set free so that the equation balances. This energy is liberated as kinetic energy of the resulting particles.

Now, two deuterium nuclei normally repel each other due to their protons. To overcome the repulsion and get them together, we have to supply energy from an external source. Fortunately, the energy we must supply to initiate the fusion is less than the $+B$ we get out of it. Hence it is practical to fuse two hydrogens. However, it is foolish to fuse two large nuclei. The repelling force due to the many protons is so great that the energy necessary to cause fusion is more than the fusion could yield.

THE NUCLEUS

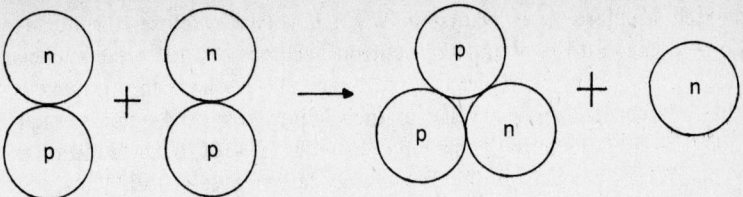

Fig. 14-2 Fusion. Two deuterium nuclei join to produce a three-particle nucleus plus an energetic neutron.

Fission

Fission is the splitting of large nuclei. It, too, can produce energy. When large nuclei split, they fly apart with tremendous energy. Let us see how we can split them.

Imagine taking a large nucleus and disturbing it so that the resulting shape is elongated. This is shown in figure 14-3, somewhat exaggerated. An elongated nucleus tends to fly apart. This can be understood by considering the elongated nucleus as two proton-filled bulges which repel each other. The only thing that keeps them from flying apart is the binding energy associated with the contact area of the narrow neck between them. If the nucleus is greatly disturbed, the contact area is reduced enough so that the binding energy becomes less than that required to keep the two parts from separating. Then the nucleus splits.

How do we disturb the nucleus in the first place? We bombard it with a neutron. A neutron is not repelled by a nucleus so it can ap-

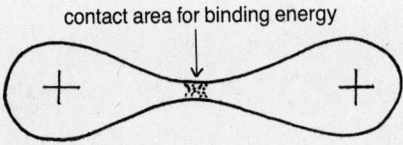

Fig. 14-3 Deformed nucleus. The repelling force of the protons can overcome the reduced binding force.

proach as close as it wants to. When it gets very close, the nuclear force comes into play and the neutron becomes bound to the nucleus. As in our fusion example, the more particles we have in the nucleus, the greater the negative value of the binding energy. Assume we start out with a nucleus that has a binding energy of $-100B$. The addition of a neutron will make the binding energy larger, say, $-105B$. Thus, $+5B$ units of energy must be liberated in order to make the energy balance. This energy goes into kinetic energy of vibration of the particles of the nucleus. It is this vibration, if severe enough, that disturbs and elongates the nucleus.

The fission of a large nucleus also liberates neutrons. These neutrons can bombard other nuclei, causing fission in them and releasing still more neutrons, etc. This process is called a *chain reaction*. These neutrons can cause a large number of fissions in a very short time. This is what happens when an *atom bomb* explodes. Not all chain reactions lead to explosions. In *atomic power plants,* the chain reaction is regulated by *control rods*. These are pieces of material which absorb neutrons, thus preventing them from causing fission. Thus, the power plant operator can decide essentially how many fissions are needed and interfere accordingly. It should be noted that in both chain reactions, we are assuming there are enough nuclei for the many neutrons to bombard. If not, the reaction does not get very far. The minimum mass needed to keep a chain reaction going is called the *critical mass*.

PART THREE

ASTRONOMY

15
Earth, Sun, and Moon

In this chapter, we will divide our attention between two areas: phenomena relating to the motion of the earth in its orbit, and the properties and behavior of the sun and moon.

THE EARTH IN ORBIT

Direction

The earth's axis intersects the surface at two points called the *north pole* and the *south pole*. Their positions are related to the direction of the earth's rotation as indicated in figure 15-1. Having established the poles and the direction of rotation, we can uniquely define four directions at any point on the earth's surface.

1) *North* (*south*) is the direction of the shortest path to the north (south) pole.

2) *East* (*west*) is the direction along (opposite) the earth's rotation.

The Earth's Coordinate System

The location of a point on the earth's surface is accomplished with a series of crisscrossing circles.

1) A *great circle* is a circle on the earth's surface whose center is at the earth's center. An infinite number of great circles can be drawn.

2) The *equator* is the one great circle whose plane is perpendicular to the earth's axis. Note that this plane bisects the axis.

3) A *parallel* is a circle that is parallel to the equator.

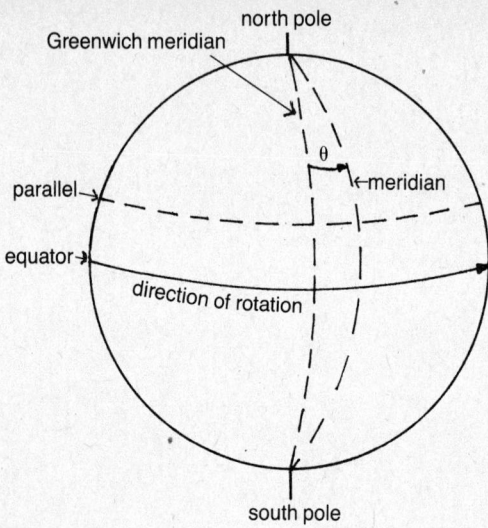

Fig. 15–1 The earth's coordinate system

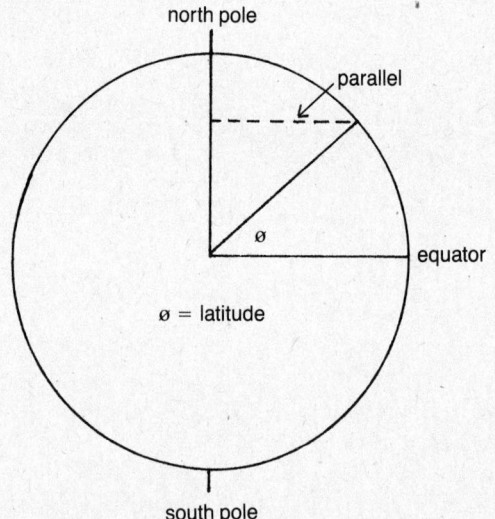

Fig. 15–2 Cross section of the earth showing how latitude is calculated

EARTH, SUN, AND MOON

4) *Latitude* is the angle which specifies the location of a parallel. This is the angle ϕ in figure 15-2, which shows a cross section of the earth.

5) A *meridian* is a circle that passes through both poles. All meridians are great circles.

6) *Longitude* is the angle θ (figure 15-1) which specifies the location of a meridian. It is the number of degrees east or west of the meridian passing through Greenwich, England. Denver, for instance, is 105° west of Greenwich. Its longitude is therefore 105°W.

Time and Day

1) A *day* is the amount of time it takes for a given object in the sky to reappear in the same position. A *solar day* is the time between successive appearances of the sun over a given meridian. A *sidereal day* is the time between successive appearances of a distant star over a given meridian.

The difference between the two can be seen with the aid of figure 15-3. The earth is shown in two positions in its orbit around the sun. Assume that the earth were to remain at position A. Then light from the sun would reappear overhead at P after one rotation of the earth about its axis. Similarly, light from the star would reappear overhead at P'. We would conclude that the solar and sidereal day are the same length. But after one day, the earth moves to position B. The distant star is so far away that its light appears to come from the same direction whether the earth is at A or B. However, the sun is so much closer to the earth than the star that its light comes from a distinctly different direction on the two days. While the light from the star still reappears overhead at P' after one revolution, light from the sun does not reappear overhead at P. From the diagram, we can see that the earth must rotate an extra amount α in order for the sun to reappear overhead. This extra rotation, which is required because of the movement of the earth in its orbit, accounts for the difference in duration between the two types of days.

The earth does not move the same distance in its orbit from day to day. Hence, α changes, along with the length of the solar day. If we average all the solar days over a year, we find that the resulting *mean solar day* is about four minutes longer than the sidereal day.

2) The *time of day* depends on the relative location of the sun and a particular meridian. *Twelve o'clock noon* is the time at a particular meridian when the sun is overhead. For example, in figure 15-3, when the

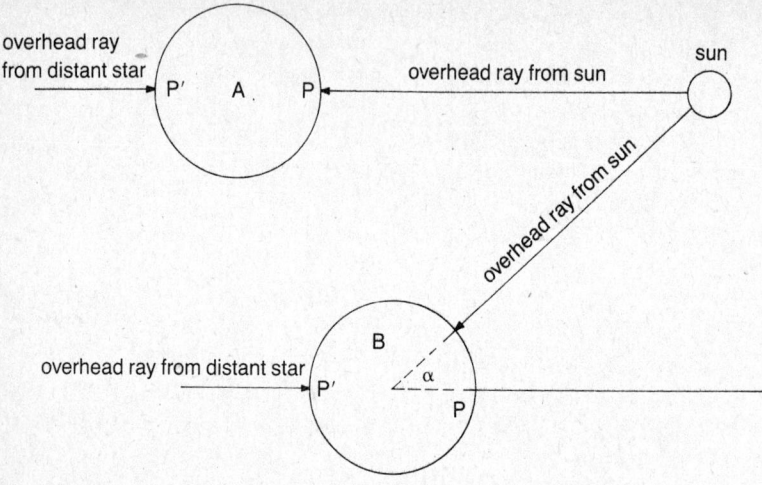

Fig. 15-3 Solar versus sidereal day. The view is from space, looking down at the north pole.

earth is at A, it is noon at point P. A day begins twelve hours before noon and ends twelve hours after noon. Both of these times are called *12 o'clock midnight*. The hours before noon are abbreviated A.M.; those after noon are abbreviated P.M. When the time at a given location passes midnight, the date is advanced by one.

3) A *time zone* is a region 15° wide in longitude such that each meridian therein is considered to be at the same time. There are 24 such zones, one for each hour. Figure 15-4 shows the four zones appropriate to the United States. The longitudes shown are the centers of zones that extend 7.5° on each side. If it is noon at 75°W, then it is noon in the entire zone, even though the sun is technically overhead just at the 75° longitude. Also, the time in each succeeding zone to the west is one hour earlier; to the east, one hour later.

4) The *international dateline* is the name given to the 180° meridian and separates regions where the dates are one day apart. The earlier

EARTH, SUN, AND MOON

date is to the east of the line. The reason for the dateline can be understood from the following scenario: Imagine that it is noon in New York on January 10. You fly west around a latitude circle with precisely the speed needed to keep the sun overhead. By definition, it is always noon on the plane, in agreement with local time. When you get back to New York 24 hours later, your time agrees with local time. However, you think it is still January 10 because you did not pass through midnight. Obviously, your date is wrong. To remedy this, it is agreed that whenever the dateline is passed going west, the date is advanced by one. Similarly, when one goes east across the dateline, the date is set back by one. If this had been done on the flight from New York, there would have been no discrepancy regarding the date. Because of the westward crossing of the dateline, the date would have been set to January 11.

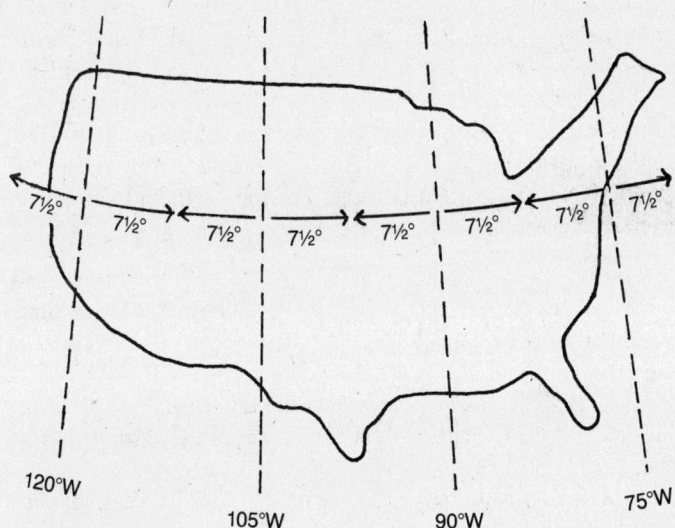

Fig. 15-4 Time zones in the United States

The Earth's Tilt and its Consequences

The earth's axis makes an angle of 23.5° with respect to the direction perpendicular to the orbit plane. This tilt is maintained throughout the orbit. Figure 15-5 shows the orientation of the axis with respect to the orbit.

Figure 15-6 is a more detailed view of the earth showing how the sun's rays are received. Half the earth is illuminated, while half is in darkness (shaded portion). The incident rays from the sun are essentially parallel to each other at all latitudes, mainly because of the large earth-sun distance. However, at only one latitude is the incident ray perpendicular to the earth's surface. We call this ray the *vertical ray*. All other rays are not perpendicular and are called *oblique rays*.

A vertical ray is inherently more intense than an oblique one. The reason for this is seen in figure 15-7, which shows a light beam of a given width. When the incident beam is perpendicular, it intercepts an area on the surface equal to S. When it is oblique, the same beam intercepts an area equal to S'. Clearly, $S' > S$. Hence, a given beam of light is more concentrated when it is perpendicular.

A careful look at figure 15-5 will reveal that the vertical ray, which is on a line with the orbit plane, is above the equator at A and below it at B, half a revolution later. As the earth makes one revolution about the sun, the vertical ray makes a trip from 23.5°N to 23.5°S and back again.

1) The *seasons* are a direct consequence of the tilt of the earth. They are simply periods of time whose beginnings are marked by the arrival of the vertical ray at a certain latitude. For the northern hemisphere, the seasons are determined as follows:

a) *Summer* begins when the vertical ray is at 23.5°N. This is called the *summer solstice* and occurs about June 21. At this time, latitudes north of 23.5° receive sunlight whose rays are closer to the vertical than at any other time.

b) *Fall* begins when the vertical ray crosses the equator going south. This occurs about September 21 and is called the *autumnal equinox*.

c) *Winter* begins when the vertical ray is at 23.5°S. Incident rays are the most oblique all year on this date. This occurs about December 21 and is called the *winter solstice*.

d) *Spring* begins when the vertical ray crosses the equator going north. This occurs about March 21 and is called the *vernal equinox*.

EARTH, SUN, AND MOON 177

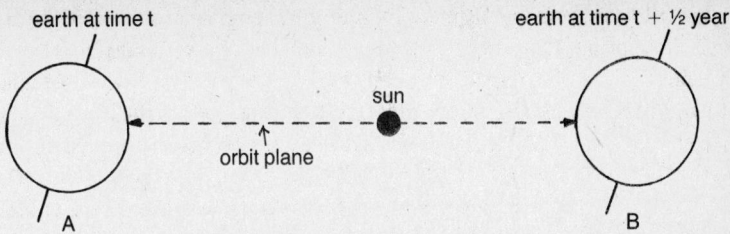

Fig. 15–5 Tilt of the earth's axis

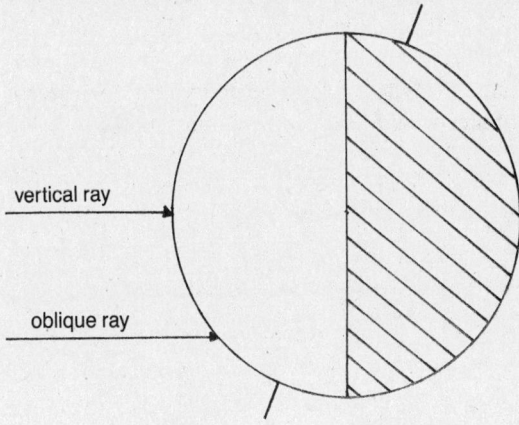

Fig. 15–6 Vertical versus oblique rays from the sun

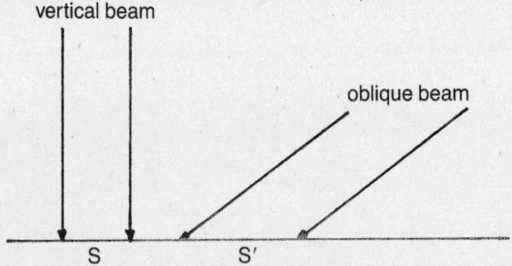

Fig. 15–7 Areas intercepted by equal vertical and oblique beams of light. The oblique beam is spread out over a wider area.

178   ASTRONOMY

For the southern hemisphere, winter and summer are reversed, as are fall and spring. One should keep in mind that the vertical ray is never found outside the region between 23.5°N and S. The lower latitudes therefore receive significantly more radiant energy than the higher ones. When we study meteorology, we will see that this fact has a profound effect on the behavior of the atmosphere.

2) The *length of daylight* varies with latitude because of the earth's tilt. In figure 15-8, the line EQ represents the path of a point on the equator as the earth rotates. Note that half of this line is in the illuminated part, while half is in the dark part. From this we see that at the equator, there is an equal amount of daylight and darkness every day of the year. Note that AB is in the illuminated zone for a longer time than in the dark zone. Also, CD is always in light, while FG is always in darkness. Clearly, the amount of daylight is latitude-dependent. If the earth were not tilted, all latitudes would have equal amounts of daylight and darkness.

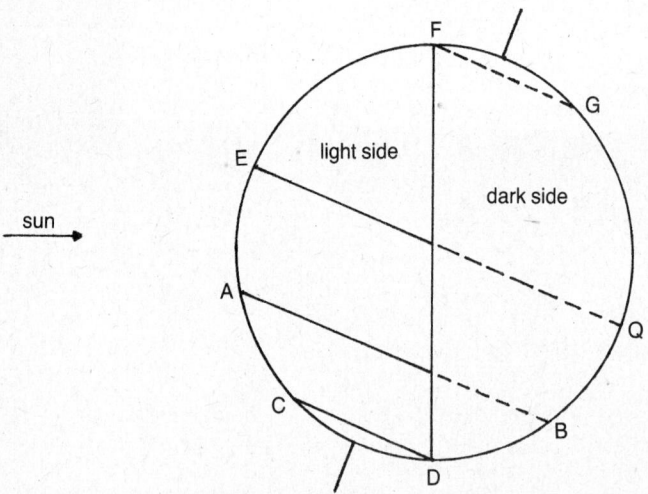

Fig. 15–8   Length of daylight at various latitudes

EARTH, SUN, AND MOON

THE SUN

The sun is a star whose proximity to earth is extremely important to us. It is the source of virtually all energy on earth. Also, unlike other stars whose image in a telescope remains a pinpoint, the details of the sun's surface are observable to scientists.

Structure of the Sun

It is convenient to divide the sun into four layers: the interior, the photosphere, the chromosphere, and the corona.

1) The *interior* comprises most of the sun's mass. Here, hydrogen is converted to helium by nuclear fusion. Four hydrogen nuclei combine to form a helium nucleus. Energy is liberated because the atomic mass of helium is about 0.83 amu less than the atomic mass of the four hydrogen nuclei. The mass difference is converted to energy according to the relation $E = mc^2$. It is estimated that fusion has been going on for about 10^8 years and could go on for another 10^{10} years.

2) The *photosphere* lies just above the interior and is the region from which most of the sunlight radiates. The photosphere is characterized by *granules*. These are regions about 10^3 km wide where hot gases rise, spread out horizontally, cool, and sink. The hot central area of each region is brighter than its cooler surroundings, resulting in a grainy appearance.

3) The *chromosphere* lies above the photosphere and is responsible for the reddish color of the sun. The color is due to the red lines of the hydrogen spectrum. Thin wisps of gas called *spicules* extend up from the top of the chromosphere into the overlying layer.

4) The *corona* is the outermost layer of the sun. It consists of ionized gases extending about 10^6 km above the surface of the sun. The *solar wind*, which consists of fast-moving electrons and protons that have escaped the sun's gravity, comes from the corona. These particles travel through space at high speed, often coming in contact with various bodies of the solar system. For instance, the surface of the moon is affected by the bombardment of the solar wind. The surface of the earth, however, is not; the earth's magnetic field diverts the particles and prevents them from reaching the surface (although they do interact with the atmosphere).

Solar Activity

1) *Sunspots* are areas of the surface which appear dark because they are at a relatively low temperature compared to their surroundings. The number of sunspots varies in time, reaching a maximum every eleven years. Also, as the number of spots increases, they tend to form closer to the sun's equator.

2) *Solar rotation* refers to the rotation of the sun about its own axis. This rotation is not uniform. Observations of sunspots indicate that it rotates fastest at the equator. This behavior would be impossible if the sun were solid instead of gaseous.

3) *Prominences* are great eruptions of gas at the edge of the sun which can reach outward for tens of thousands of kilometers. There is a magnetic field associated with them; the arched shape of many prominences results when the erupted gas falls back to the chromosphere along the magnetic lines of force. Prominences are usually viewed when most of the sun's disk is blocked out and just the edge is visible. Sometimes, they can be seen against the main body of the sun. Then they are called *filaments* and appear as dark streaks.

4) *Solar flares* are explosions associated with sunspots. They send out tremendous bursts of X-rays and charged particles which cause increases in the solar wind.

5) *Auroras* are spectacular colored displays near the poles of the earth. They result when charged particles from the sun, guided to the poles by the earth's magnetic field, interact with the upper atmosphere. Electrons in the atmospheric gases jump to higher energy levels as a result of these interactions, then emit radiation when they drop back to lower energy levels. The emitted radiation is visible as the aurora.

THE MOON

The moon is the only natural satellite of the earth. Although many planets have satellites, only the earth has one whose size is a significant fraction of the main planet. The moon's periodic motion results in the so-called phases and, occasionally, eclipses. Its gravitational pull is important in producing tides. All of these phenomena will be explored in this section.

EARTH, SUN, AND MOON

The Moon's Orbit

The actual path of the moon through space is quite complicated, since its motion around the earth is combined with the earth's motion around the sun. For most purposes, we can leave out the complications and think of the moon's motion as simply a circle around the earth. When discussing the tides, however, we must look at the orbit more closely.

1) The *phases* of the moon are different positions of the moon with respect to the earth and sun. Figure 15-9 shows what this means; it

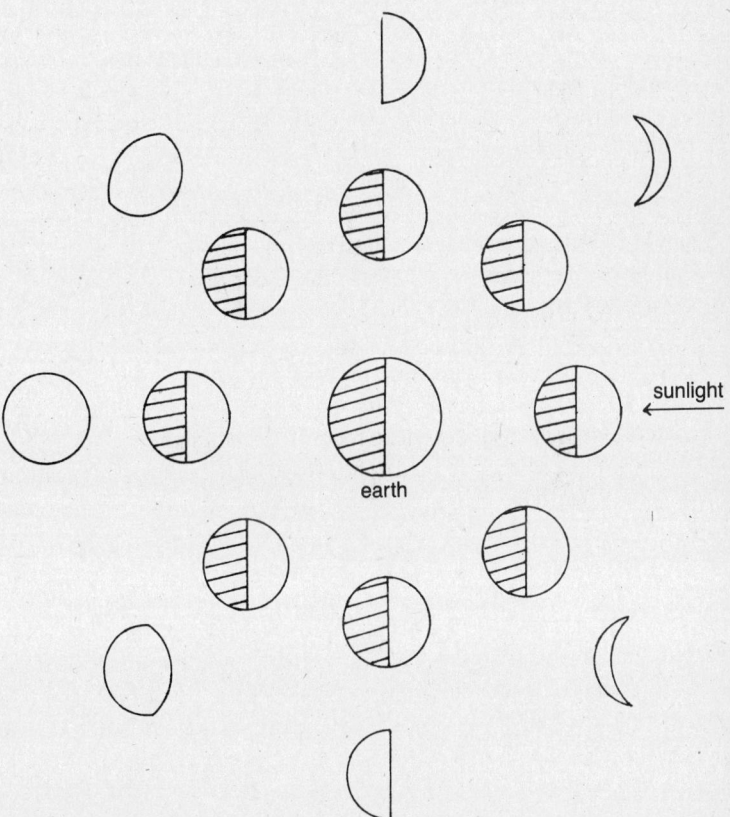

Fig. 15–9 Phases of the moon. The inner circles show the moon in different positions. The outer figures show the view from the earth directly below.

should be studied carefully. Shown here are eight different positions of the moon. Next to each position is the view of the moon from a meridian directly underneath it. Note that although half of the moon is always illuminated, what you see is a mix of dark and light areas that depends on your position.

2) The *synodic month* is the time between successive new-moon phases, 29.5 days. The *sidereal month* is the time for the moon to make one complete revolution around the earth, 27.3 days. The difference between the two is illustrated in figure 15-10. This difference is similar in origin to the difference between the solar and sidereal day. At the start, a *new* moon, a distant star, and the earth are all lined up. After 27.3 days, the earth, moon, and distant star are lined up again. The moon has therefore completed a full revolution around the earth. As we can see from the diagram, some extra time must elapse before an earthbound viewer sees the next *new* moon. Hence, the synodic month takes longer to complete than the sidereal month.

3) A *solar eclipse* occurs when the moon casts a shadow on the surface of the earth. A *lunar eclipse* occurs when the earth's shadow falls on the moon. Both kinds of eclipse are illustrated in figure 15-11.

Because of the geometry involved, the moon's shadow consists of a dark central part, called the *umbra,* and a light outer part, called the *penumbra.* Viewers entirely in the path of the umbra experience a *total eclipse.* They see only the chromosphere and corona. Viewers in the penumbra see a *partial eclipse:* a slice of the sun is always visible.

In a total lunar eclipse, the entire moon is in the earth's shadow. However, the moon can still be seen. The red portion of the sun's spectrum is refracted slightly by the earth's atmosphere and hits the moon, giving it a copper hue. In a *partial lunar eclipse,* only part of the moon is in the shadow.

Eclipses are rare even though the moon is often in front of or behind the earth. This is because the orbit of the moon around the earth is not in the same plane as the orbit of the earth around the sun. Most of the time, for instance, the moon's passage between the earth and sun is in a position such as P or P' in figure 15-11. Hence, a shadow is not cast on the earth.

EARTH, SUN, AND MOON

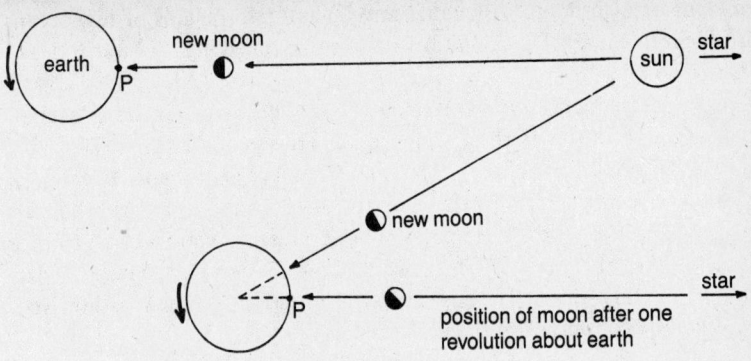

Fig. 15–10 Synodic versus sidereal month

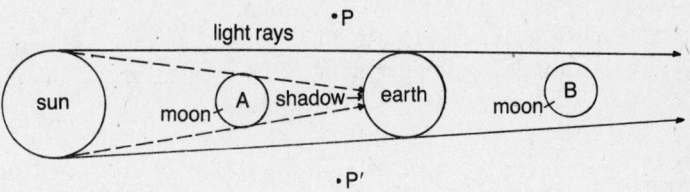

Fig. 15–11 Relative positions of the sun, earth, and moon for solar (A) and lunar (B) eclipses

Tides

A *tide* is a periodic change in the height of the ocean surface in a given location. When looked at in cross section, the ocean appears as a thin skin of water with two bulges as shown in figure 15-12. The bulges are always lined up with one toward the moon and one away from the moon. As the earth rotates, a given point on the surface, *P*, passes through two regions of high ocean surface (*A* and *C*) and two regions of low ocean surface (*B* and *D*). The condition at *A* or *C* is called *high tide;* at *B* or *D*, it is called *low tide*.

1) Tide-producing forces. We know that the moon exerts a gravitational force on the earth. This force pulls the solid earth and the water toward the moon. The solid earth, of course, keeps its shape. The water, however, tends to flow to the side of the earth closest to the moon. If the gravitational force were the only one present, water would tend to pile up at *A* but not *C*. There would be only one high tide. Why are there *two* high tides? The answer to this question is somewhat involved and will be considered in the next few paragraphs.

Strictly speaking, the moon does not revolve around the earth. Approximately once a month, the earth and moon revolve around a point between their centers (figure 15-13). This point is called the *barycenter*. The combined motion can be visualized by imagining that the dashed line in figure 15-13 simply rotates around the barycenter. In reality, the monthly motion of the earth is combined with a daily rotation about the polar axis and a yearly revolution around the sun. However, the latter two motions have nothing to do with the tides and must be subtracted out. Only the resulting motion, which is called *revolution without rotation,* is involved.

It is extremely important that you understand revolution without rotation. Fortunately, you can demonstrate it for yourself. Take a square or rectangular piece of cardboard and locate the center. Poke one pencil through the center and another anyplace else. Orient the cardboard over a clean sheet of paper so that the sides of the paper and cardboard are parallel. Now, keeping them parallel, move the cardboard over the paper so that the pencils trace out circles. What happens is this: The center of the cardboard (represented by the pencil through it) moves in a circular path. So do all other points (represented by the other, arbitrarily placed pencil). However, the other points do not rotate about the center of the cardboard. We say that the cardboard executes revolution without rotation. Note that the circles traced out have the same radius. This demonstrates that all points on the cardboard execute the same amount of revolution without rotation.

The earth is analogous to the cardboard. Try to imagine the earth in figure 15-14 moving around the barycenter but not rotating about its own axis. (Just keep line PC_e horizontal.) The result is revolution with-

EARTH, SUN, AND MOON

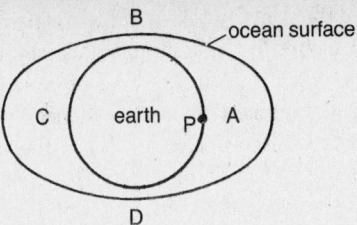

Fig. 15–12 Tides

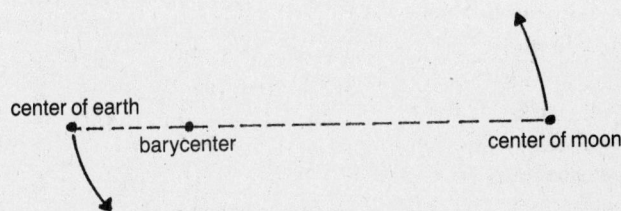

Fig. 15–13 The earth–moon system. The center of each body goes around the barycenter once a month.

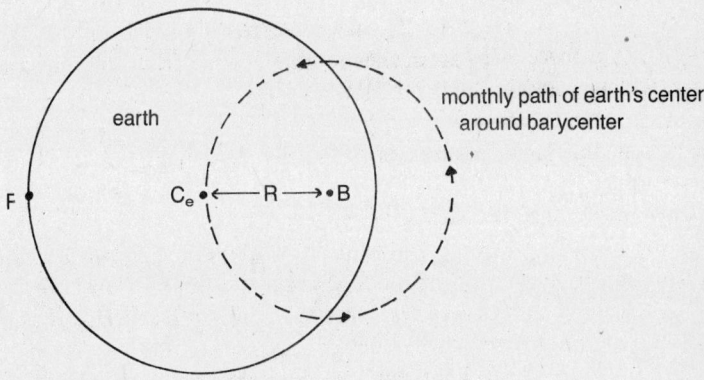

Fig. 15–14 Revolution without rotation. Imagine that the earth moves in the indicated path so that a line from P to C_e is always horizontal in the diagram.

out rotation. The dashed circle shows the path that the center of the earth traces out in a month. You should understand that every other point on the earth would trace out a circle of equal radius, just as the two pencils traced out equal-size circles. Hence, every point on the earth executes the same amount of revolution without rotation.

Let's consider the consequences of this motion. The most convenient way to do this is to consider a frame of reference rotating about the barycenter. From our discussion of rotating frames in chapter 1, we know that a particle of water in this frame feels a centrifugal force. Since all particles of water execute exactly the same amount of revolution without rotation, we can conclude that all particles of water must feel the same centrifugal force. But in what direction is this centrifugal force? We know that centrifugal force is always in a direction radially outward from the center of rotation. Referring back to figure 15-13, we can see that this is the same as the direction from the moon to the earth. Hence, all particles of water feel an equal centrifugal force in a direction directly away from the moon.

Now we are ready to see why there are two tides. The force that produces the tides is the vector sum of the gravitational and centrifugal forces that act on the water. Figure 15-15 illustrates this. On the side nearest the moon, the gravitational attraction is slightly greater than the centrifugal force. Thus, there is a net force toward the moon. However, on the side away from the moon, the gravitational attraction is weaker than the centrifugal force. This is because gravitational force weakens with distance, while the centrifugal force remains constant. Hence, there is a net force away from the moon. These net forces produce the bulges which represent the two high tides.

2) Solar effects. The *sun* also produces tides; because of its great distance from the earth, solar tides are smaller in magnitude than lunar tides.

a) In a *neap tide,* the difference between high and low tide is a minimum. It results when the sun, moon, and earth line up as shown in figure 15-16. Tides produced by the moon are high at A and C, low at B and D. The smaller tides produced by the sun are high at B and D, low at A and C. The result is that high tides are not as high as they would be without the sun. Similarly, low tides are not as low.

b) In a *spring tide,* the sun intensifies the tide-producing force due to the positioning shown in figure 15-17. In this case, both sun and moon tend to produce high tides at A and C, low tides at B and D. Thus, the high tide is higher than if there were no sun; similarly, low tide is lower.

EARTH, SUN, AND MOON

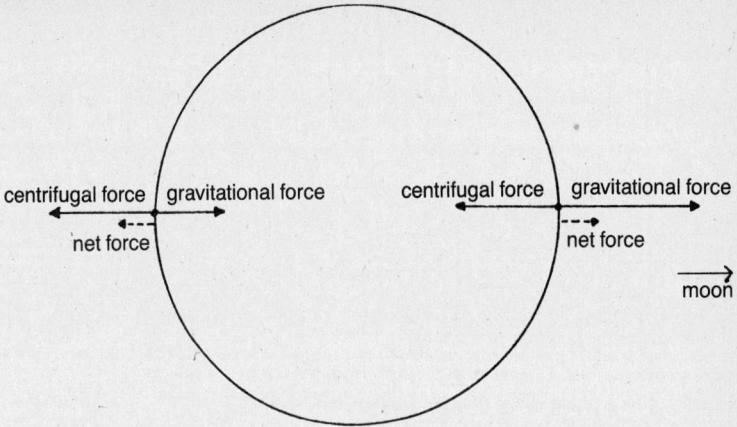

Fig. 15–15 Tide-producing forces

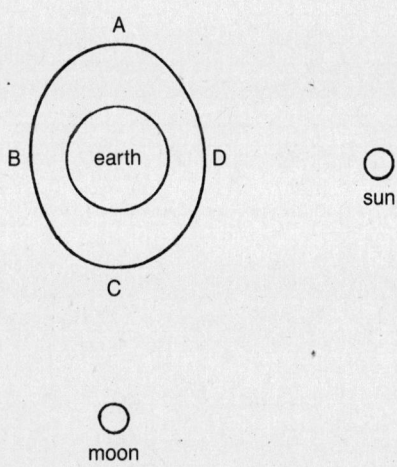

Fig. 15–16 Neap tides

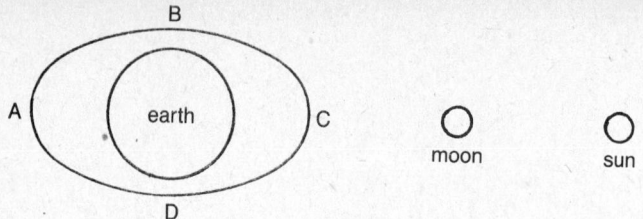

Fig. 15-17 Spring tides

The Lunar Surface

The lunar surface is known as *regolith*. This is nothing more than a layer of debris that is the result of billions of years of bombardment by various objects from space. Although this activity ceased about 10^9 years ago, the moon's surface has hardly changed since then. This is because the moon lacks water and atmosphere; there is no erosion or weathering as there is on earth.

When viewed through a telescope, there are two main surface features that stand out.

1) *Craters* are depressions of varying sizes on the moon's surface. A typical large crater is about 250 km in diameter and often overlaps other craters. Craters are formed when rapidly moving objects in space collide with the moon's surface. These impacts can produce craters fifty times the diameter of the crashing object. The process is similar to a rock falling into water. Material under the impact area is pushed out of the cavity created by the impact. In the case of the crater, most of the material lands near the cavity, forming a rim.

2) *Maria*, which is Latin for "seas," are smooth, low-lying areas. They are huge impact craters that have been flooded by lava flow. It is commonly believed that the impact that produced a given mare caused widespread breakage in the surrounding crust. Often, a lava flow covered the original crater plus its surroundings, producing a vast smooth area. The Sea of Tranquility, where the first moon landing took place, is such an area.

The centers of circular-shaped maria are the sites for *mascons* (*mas*s *con*centrations), regions where the density is greater than usual. As a result, the gravitational pull over these regions is higher than normal. This has an effect on satellites in lunar orbit, whose speed is increased over the mascons.

16
The Solar System

The solar system consists of the sun, moon, planets, thirty-five moons of other planets, and large numbers of bodies called asteroids, comets, and meteoroids.

PLANETARY MOTION AND KEPLER'S LAWS

Theories of Planetary Motion

1) The *geocentric theory* states that everything in the heavens revolves around the earth. Although this theory is wrong, the ancients were not unreasonable in believing it. From the earth, it appears that the heavens move and the earth stands still. Also, ancient astronomers, notably Ptolemy, worked out a cumbersome system based on geocentrism that enabled them to explain planetary motion to their satisfaction.

2) The *heliocentric theory* states that the planets, including the earth, revolve around the sun. It was first proposed over two thousand years ago by the astronomer *Aristarchus* (about 310-230 B.C.), but was rejected largely due to the influence of *Aristotle* (384-322 B.C.). In the early 1500s, *Nicholas Copernicus* (1473-1543) revived the idea. The first observational evidence for the heliocentric theory was provided by *Galileo* (1564-1642). He discovered four moons that orbited Jupiter. This proved that the earth is not the only center of rotation. He also discovered that Venus has phases similar to the moon's. In the geocentric theory, Venus would have only a crescent phase. Unfortunately for Galileo, the church regarded his findings as heresy; he spent the remaining years of his life under house arrest.

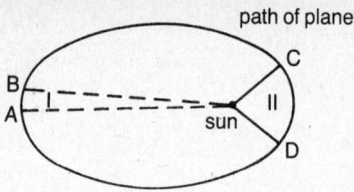

Fig. 16-1 Kepler's Second Law

Kepler's Laws

Johannes Kepler (1571-1630) formulated mathematical relationships (Kepler's Laws) which supported the heliocentric theory. He based his work on a lifetime of careful observations by an astronomer named *Tycho Brahe* (1546-1601).

1) *Kepler's First Law* states that the planets orbit the sun in an elliptical path, with the sun at one focus. (The ellipse is discussed in the appendix.)

2) The *Second Law* states that the radius vector (the imaginary line joining the planet to the sun) sweeps out equal areas in equal times. Figure 16-1 shows what this means. In region I, the planet covers the distance *AB* in a certain time. In region II, it covers *CD* in the same time. Since both regions must have the same area, *CD* must be greater than *AB* in order to make up for the smaller radius in region II. Hence, the planet must cover a greater distance in the same time (i.e., move faster) when it is closer to the sun.

3) The *Third Law* relates the period of revolution, *T*, of a planet with the semimajor axis (the average distance of the planet from the sun), *D*:

$$T^2/D^3 = K$$

Here, K is a constant that is the same number for all the planets. Note that the first two laws state that the same *type* of motion underlies the behavior of all the planets. The Third Law goes even further by stating that a particular *numerical* value is the same for all planets.

Kepler's work established mathematical patterns implicit in the huge mass of data he examined. He showed that the heliocentric theory could account for the motion of the planets, including details that were troublesome for the geocentric theory. Furthermore, the mechanics of ature of planetary motion are much simpler than in the geocentric theory. However, he did not demonstrate *why* the planets behaved this way. Such an explanation had to wait for Newton (1642-1727). Using calculus, which he invented, and the laws of physics, Newton showed that planetary motion is a direct consequence of the nature of the gravitational force.

THE CELESTIAL SPHERE

For the purpose of locating heavenly bodies, it is convenient to regard the earth as a stationary body located at the center of an imaginary sphere called the *celestial sphere*. The stars are assumed fixed in position on the celestial sphere, which executes an apparent rotation about an axis coincident with the earth's axis.

The Equatorial Coordinate System

The *equatorial coordinate system* is a system of latitude and longitude circles used to locate a point on the celestial sphere. These are just like the earth's latitude and longitude circles, except that they are inscribed on the celestial sphere. Figure 16-2 illustrates the details of this system.

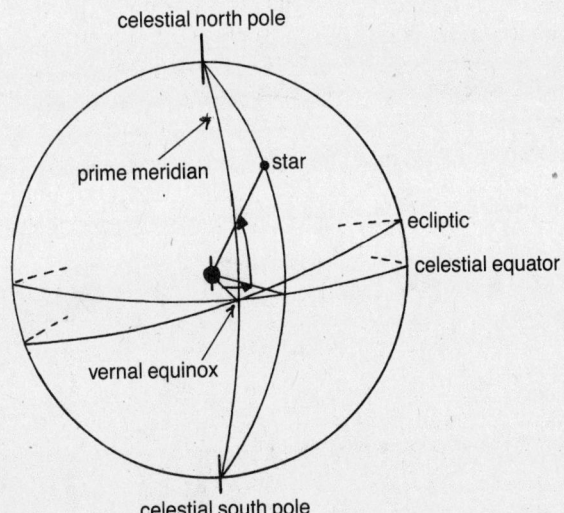

Fig. 16–2 Equatorial coordinate system. Horizontal arrow indicates right ascension; vertical arrow indicates declination.

1) The *celestial poles* are points where the earth's axis, when extended, intercepts the celestial sphere. The north and south celestial poles are directly above their respective earth counterparts.

2) The *celestial equator* is a circle on the celestial sphere which is an extension of the earth's equator.

3) The *ecliptic* is the path of the sun on the celestial sphere. The ecliptic intersects the celestial equator at two points. At these intersections, the sun is directly over the equator. Hence, these two points are the equinoxes.

4) The *celestial prime meridian* is the longitude which is drawn through the vernal equinox.

5) *Right ascension* is the angular measure of celestial longitude, usually measured in hours east of the celestial prime meridian. The use of hours as the units of right ascension is convenient because it takes 24 hours for a star to return to its initial right ascension.

6) *Declination* is the angular measure of celestial latitude and is measured in degrees north or south of the celestial equator.

Constellations

A *constellation* is a group of stars that is recognizable as a distinctive pattern, especially if you have a good imagination. Of particular interest as far as constellations are concerned is a section of the sky called the *zodiac* (figure 16-3). This is a path 8° above and below the ecliptic that

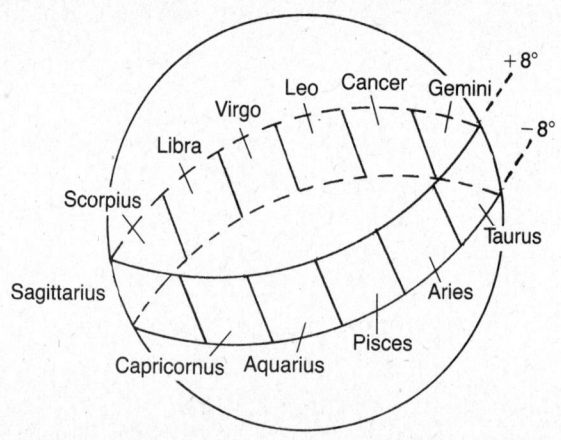

Fig. 16-3 Constellations of the zodiac

THE SOLAR SYSTEM

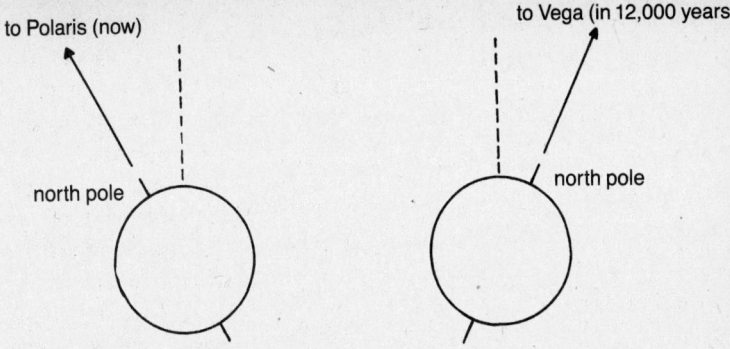

Fig. 16-4 Precession of the earth's axis

encircles the celestial sphere. It is divided into twelve equal sections, each with its own identifying constellation. These are Aries, Pisces, Aquarius, Capricornus, Sagittarius, Scorpius, Libra, Virgo, Leo, Cancer, Gemini, and Taurus. The zodiac is the background for the sun, moon, and planets in their apparent motion around the earth. Hence, it is a kind of reference frame for locating a body of the solar system. For instance, the vernal equinox occurs when the sun is in Pisces. Similarly, each successive new moon is found in one of the twelve constellations, in the order listed above.

The earth's axis undergoes a wobbling motion called *precession*. This is the same motion seen in a spinning toy top. After a few seconds, the axis of the top, which leans somewhat, slowly changes its direction. We can understand this by looking at the earth in figure 16-4. At present, the earth's axis is oriented toward the star Polaris. In another 12,000 years, the axis will point toward Vega. The sky will look quite different; the vernal equinox will no longer be in Pisces. A look back at figure 16-2 shows why. As the axis precesses, so does the celestial equator. Hence, the celestial equator will intersect the ecliptic in a different constellation.

THE PLANETS

With the exception of Pluto, whose exact nature is not well known, the planets can be put into two categories: *terrestrial*, meaning earthlike, and *Jovian*, meaning Jupiterlike.

Terrestrial versus Jovian Planets

In general, the two classes of planets differ in mass, atmosphere, composition, and rotation rate.

1) The *mass* of a typical Jovian planet is much larger than that of any terrestrial planet.

2) The *atmosphere* of terrestrial planets is thin, while that of the Jovian planets is dense. The amount of gas that can be held near the surface of a planet depends mostly on the gravitational attraction, which in turn depends on the mass.

3) The *composition* of the terrestrial planets is largely dense rock and metals. A representative figure for the density is 5 g/cm^3. The density of the Jovian planets is much lower (1.5 g/cm^3) because they consist mainly of hydrogen and helium, plus some frozen ammonia and methane.

4) The *rotation rate* is faster for the Jovian planets.

Terrestrial Planets

1) *Mercury* is the smallest planet and the one closest to the sun. Its orbital period is only 88 days, but it takes 59 days to rotate about its own axis. Hence, both days and nights are long enough to permit large extremes in temperature. The range is -140°C to 315°C. As a result, there is probably no life on Mercury.

2) *Venus* is characterized by a very thick layer of clouds and an atmosphere that is mostly carbon dioxide. As on earth, the CO_2 acts as an insulator, causing the surface temperature to rise to levels much higher (410°C) than if no atmosphere were present. Also, the surface pressure is about 90 times that of the earth. These conditions not only make life impossible but quickly destroy the space probes that land on its surface. The Soviet probe Venera, for instance, lasted only one hour.

The surface of Venus is obscured by the clouds. However, the Venera missions noted the presence of both angular and smooth rocks. Radar studies have also detected moonlike craters.

3) *Mars* has a rotation rate and a tilt very similar to that of the earth. Because of the tilt, Mars has seasons. However, its period of revolution is about 50 percent longer than earth's, so the seasons are longer. They are also cooler because Mars is farther from the sun.

The most notable feature of Mars is the ice which covers the poles. These *polar ice caps* expand toward the equator in the winter and re-

THE SOLAR SYSTEM

cede in the summer. From information gathered by Mariner probes and Viking landers, we now know that these caps are covered by a layer of frozen CO_2. It is the layer of CO_2 that grows and recedes with the seasons.

The surface of the southern hemisphere is characterized by moonlike craters. The northern hemisphere surface has volcanoes and lava plains. In addition, there is evidence of earthquakes, past volcanic activity, deep canyons, and water erosion. This last feature is puzzling because there is no liquid water on Mars. As yet, there is no satisfactory theory to account for the liquid water needed for erosion.

The Martian atmosphere is much less dense than that of the earth. Despite this, fierce dust storms take place. They are responsible for sand dunes, the filling of craters, and, possibly, color changes that are visible through telescopes.

Jovian Planets

1) *Jupiter* is by far the largest planet. Through a telescope, one can see colored bands of clouds that run parallel to the equator. There is also a huge red spot which is probably a storm caught between two oppositely directed jet streams (bands of strong winds).

The exact structure of Jupiter is a matter of conjecture. We know that its atmosphere, which is largely hydrogen and helium, exerts tremendous pressure. It has been calculated that, 160 km below the clouds, the pressure is enough to liquefy hydrogen. Thus, an ocean of liquid hydrogen, whose surface is at this depth, has been postulated. It is not certain what lies below the ocean. Some studies suggest a core of rock; others suggest a slushlike material composed of frozen and liquid gases.

Jupiter has fourteen moons, four of which can be seen with a low-power telescope. These are the ones that were discovered by Galileo. Because they have short orbital periods, an amateur observer can follow the orbits easily, even tracing their shadows as they move in front of Jupiter.

2) *Saturn* is thought to have a composition similar to that of Jupiter. It is most noteworthy for its concentric rings. There are apparently thousands of these rings, according to pictures sent by the Voyager spacecraft. The rings are not solid. They consist of small bodies up to a few meters in diameter, possibly solid NH_4. The material in the inner

portion of a ring orbits faster than the material in the outer portion. This would be impossible if the ring were solid. It is likely that the rings are part of the material which originally formed Saturn itself. The Voyager spacecraft showed that Saturn's atmosphere is characterized by turbulent weather. There are jet streams that appear to flow even faster than those on Jupiter. Also, there are huge storm systems, as wide in horizontal extent as Europe and Asia.

Saturn has ten moons, all located beyond the outer ring. One of these, Titan, is the only moon in the solar system that has an atmosphere. Voyager also showed the shape of Hyperion to be that of a potato, while Tethys has a huge crater about one third the diameter of the moon itself.

3) *Uranus* was discovered accidentally in 1781 by William Herschel, for whom astronomy was a hobby. Much remains unknown about Uranus. In 1984, however, the Voyager spacecraft is expected to fly by Uranus and send back new information. Like Saturn, it has rings (at least five). Its atmosphere has hydrogen and methane, but not ammonia. The ammonia is probably in the solid state. In fact, one theory is that Uranus is mostly frozen methane and ammonia.

The axis of Uranus is tilted at an angle of 82°, making it almost parallel to the orbit plane. This is in sharp contrast to the other planets, whose tilts are much more vertical.

4) *Neptune* was discovered after deviations showed up in the calculated orbit of Uranus. Around 1845, two astronomers, Adams and Leverrier, independently assumed that the deviations were due to the gravitational force from an undiscovered planet. This led to the discovery of Neptune. In terms of structure, Neptune is believed to be a virtual twin of Uranus, although its greater distance from the sun makes it colder.

Pluto

Pluto is smaller than any of the terrestrial planets and farther away than any of the Jovian ones. Its discovery was also prompted by irregularities in the assumed orbit of Uranus, not all of which had been resolved with the discovery of Neptune. Pluto most likely consists of rock and frozen gases. The average temperature of Pluto is so low (-210°C) that any gas is probably a solid. Hence, it does not have an atmosphere.

Pluto has one moon. It is so small that it cannot be seen as such. It appears on photographic plates as a slight bulge in the planet itself.

THE SOLAR SYSTEM

OTHER SOLAR BODIES

Asteroids

Asteroids are small bodies, usually about 1 km in diameter, that orbit the sun between Jupiter and Mars. There are over 50,000 of them, but only the 1800 largest ones have had their orbits calculated.

The fact that their orbits lie between Jupiter and Mars is significant. There is a mathematical relation in astronomy called *Bode's Law* that relates the distance to the sun to a simple series of integers. If one takes the numbers 0, 3, 6, 12, 24, 48, ... adds 4 to each one, then divides by 10, the result is the distance to the sun in astronomical units (AU). One AU is the distance from the earth to the sun: 1.5×10^{11} m.

Example: Work out Bode's Law for 12.

Solution: Add 4: $12 + 4 = 16$
divide by 10: $16/10 = 1.6$

The planet Mars is located about 1.6 AU from the sun.

It turns out that there is no planet at 2.8 AU. This, however, is the location of the belt of asteroids. It has led to speculation by some that a planet located there somehow broke up into many fragments. Others suggest that a few large bodies collided and broke up into smaller ones. There is no hard evidence yet to support either theory.

Comets

Comets are bodies that consist of many rock fragments held together by frozen gases. Their elliptical paths are highly eccentric, coming from far beyond Pluto and approaching very close to the sun (sometimes within 1 AU). As they get closer to the sun, solar energy causes the frozen gases to evaporate and ionize. Most comets have three distinct features (figure 16-5):

1) The *coma* is a mass of glowing gas at the head of the comet. A coma is typically as large as Jupiter.
2) The *nucleus*, located at the center of the coma, is a glowing mass a few kilometers in diameter.
3) The *tail* consists of ionized gas and dust. The ionized gas is pushed away from the sun by the solar wind. The dust is pushed away by the pressure of the photons themselves.

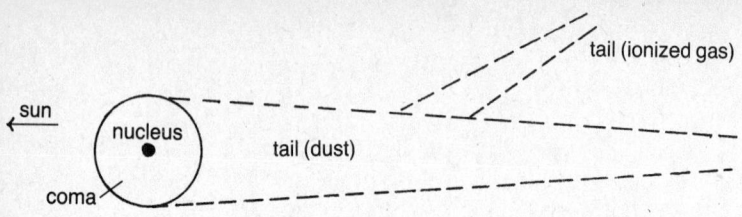

Fig. 16–5 Typical comet. The dust tail of a comet always points away from the sun.

When the comet recedes from the sun, its mass is reduced because of gas that was blown away from the coma. Hence, a comet is eventually left without any gas at all. Only the fragments orbit the sun.

In February, 1986, Halley's comet is expected to return. It was last seen in 1910 and was bright enough to be seen in daylight. Its tail was about a million miles long.

"Shooting Stars"

A *shooting star* is a popular term for any object that moves brightly and rapidly across the sky.

1) A *meteoroid* is a small particle that enters the earth's atmosphere and burns up due to friction. These particles are often the size of a grain of sand. They are so small that although we can see the glow, they burn up before they hit the surface.

2) A *micrometeorite* is a particle that is so small it cannot get up enough speed to cause friction to burn it.

3) A *meteorite* is a meteoroid that is big enough to make it all the way to the earth's surface. In Arizona, the result of a meteorite is evident in the Barringer Crater, a huge hole that is 1.2 km wide and 170 m deep.

4) A *meteor shower* is a collection of meteoroids that fall in the same direction and hit the atmosphere in relatively rapid succession.

17
The Universe

In this chapter, we will look outside the confines of our relatively small neighborhood, the solar system.

STARS

A *star* is a large body of luminous gas, held together by the force of gravity and energized from within by nuclear fusion.

Physical Properties of Stars

Information about the properties of stars comes from analyzing the radiation they emit.

1) The *apparent magnitude* of a star is a number that indicates how bright the star appears when viewed from the earth. The scale is set up so that the smaller the number the brighter the star.

Originally, the scale ran from 1 to 6. Two stars differing by 1 on this scale differ in brightness by a factor of about 2.5. For instance, a star with a magnitude of 1 is about 2.5 times brighter than a star with magnitude 2. Also, there are stars that are brighter than magnitude 1. Hence, negative magnitudes are possible. For example, the sun, the brightest object we see, has an apparent magnitude of -26.7.

Apparent magnitude does not indicate how bright a star really is. A star could appear bright because it is close to the earth. Similarly, it could appear dim because it is far away.

2) The *absolute magnitude* is the brightness a star would have if it were a standard distance from the earth. The scale enables astronomers

to compare the true brightness of stars because differences in distance are eliminated. Most stars fall in the range of −5 to +15. For instance, the absolute magnitude of the sun is actually 4.9, not the apparent −26.7.

3) The *color index* of a star is the difference in magnitude between the blue and yellow components of its radiation:

$$\text{color index} = m_b - m_y \qquad m = \text{magnitude}$$

It is a number which indicates the temperature of a star. A hot star emits more blue than yellow because most of its energy is in the short wavelength region. A cool star emits more yellow than blue because most of its energy is in the long wavelength region. Since a small m means a large amount of emitted energy, we see that for a hot star (more blue), $m_b < m_y$, so the color index is negative. For a cool star (less blue), $m_y < m_b$, so the color index is positive.

4) The *spectral class* is a collection of similar-looking absorption spectra from various stars. Spectral class is directly related to the temperature and is designated by a letter. Table 17-1 lists the seven classes with their letters, colors, and temperatures.

TABLE 17-1. SPECTRAL CLASSES

Class	Color	Temperature $(°K \times 10^3)$
O	Blue	>30
B	Blue-White	10.5–30
A	White	7.5–10.5
F	Yellow-White	6–7.5
G	Yellow	5–6
K	Orange	3.5–5
M	Red	<3.5

Distances to Stars

The distance to a star is calculated by measuring the angle called the *stellar parallax,* shown in figure 17-1. Note that a right triangle can be drawn using the sun, earth, and star as the apexes. When the length of one side and the magnitude of one angle are known, any other side or angle can be computed. Since the distance to the sun is already known, a determination of the stellar parallax enables the calculation of the distance to the star.

THE UNIVERSE

1) A *parsec* is the distance to a star that results in a parallax angle of one second (1/3600 of a degree).

2) A *light year* is the distance that light travels in one year. One parsec equals 3.26 light years.

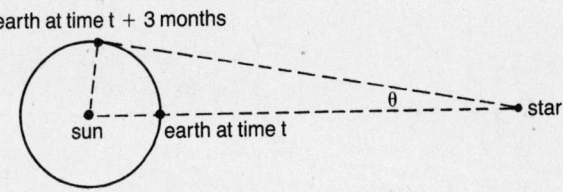

Fig. 17-1 Stellar parallax. Knowledge of the earth's radius and θ enables scientists to compute the distance to the star.

Hertzsprung-Russell Diagrams

The *Hertzsprung-Russell* (H-R) diagram is a plot of absolute magnitude (vertical scale) versus temperature (horizontal scale). A typical plot is shown in figure 17-2. Note that most stars have locations on the H-R diagram along a diagonal path called the *main sequence*.

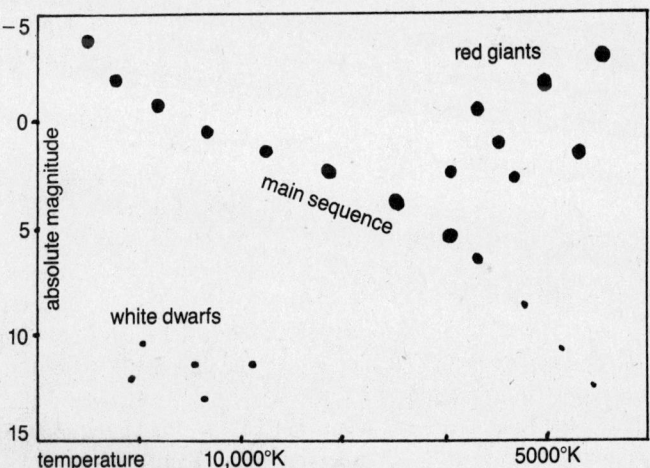

Fig. 17-2 Hertzsprung–Russell diagram

1) *Giants* and *supergiants* are stars whose positions on the H-R diagram are to the right of, and above, the main sequence. Two stars at the same temperature radiate the same energy per unit area. If one has a greater brightness, it must be due to a greater size. Hence, the stars in the upper right must be larger than stars at the same temperature but on the main sequence. These stars are at temperatures that make them appear red, so they are often called *red giants*.

2) *White dwarfs* are stars whose H-R plots are below and to the left of the main sequence. By reasoning similar to that above, they must be smaller than stars at the same temperature but located on the main sequence. They are not necessarily white, however.

Types of Stars

In addition to the giants and dwarfs described above, there are stars with some notable characteristics.

1) A *variable star* is one whose brightness fluctuates. A *pulsating variable* executes regular periodic fluctuations. A *cepheid variable* is a pulsating variable whose period is directly proportional to its absolute magnitude. By comparing absolute and apparent magnitudes, astronomers can figure out the distance to the star. An *eruptive variable* undergoes a sudden increase in brightness, then slowly reverts back to its original brightness. The increase, which can be several thousandfold, can occur within a day; the decrease generally takes about a year. The newly bright star is called a *nova*. If the brightness increases by a factor of a million, the resultant star is called a *supernova*.

2) A *binary star* is a system of two stars revolving about a common center of mass. *Visual binaries* are far enough apart to be seen with a telescope. *Spectroscopic binaries* are too close to be resolved visually. They are detected by observing slight shifts in their spectral lines. Recall that a line is due to emitted radiation of a certain frequency. If the star is in motion, there is a change in the apparent frequency due to the Doppler effect. The greater the velocity, the greater the magnitude of the change. If the star moves toward the earth, the change is toward the high frequency, which is the blue end of the spectrum. We call this a *blue shift*. If the star recedes, the change is toward the low frequency, the red end of the spectrum. This is called a *red shift*. Since the frequencies are changed, the spectral lines are shifted.

3) A *dwarf* is a small star with an extremely high (10^6 gm/cm^3) density resulting from extreme close packing of the atoms. In these atoms,

THE UNIVERSE

electrons are not in their usual orbitals; they are pushed very close to the nucleus so that the atom takes up very little space. A *white dwarf* is an example of this. A *black dwarf* is a dwarf that is nonluminous.

4) A *neutron star* is one whose atomic electrons combine with the protons to form neutrons. This material is so dense that a teaspoonful has a mass of about 10^{12} kg. A *pulsar* is a rapidly rotating neutron star which emits radiation in the form of radio pulses.

5) A *black hole* is a star that is so dense that it has enough gravitational attraction to prevent anything, including light, from escaping its surface.

6) A *star cluster* is a group of stars. A *globular cluster* is organized into a spheroidal group, while an *open cluster* has no particular shape at all.

Stellar Evolution

Stars go through a life cycle involving various stages. These are outlined below and illustrated on the H-R diagram of figure 17-3.

1) The *birth* of a star occurs when a large mass of interstellar matter contracts under the influence of gravitational attraction. Due to contraction, gravitational potential energy decreases (just as a falling object loses *PE* as it gets closer to the earth's surface). This is balanced by an

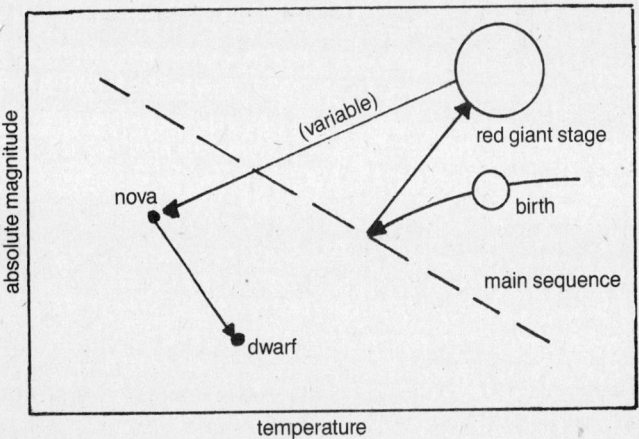

Fig. 17-3 Stellar evolution as plotted on H-R diagram

increase in thermal energy: the temperature of the star increases.

2) In the *hydrogen burning* stage, the temperature rises enough to permit the fusion of hydrogen to helium. This lasts for billions of years. During this stage, the star is on the main sequence of the H-R diagram.

3) The *red giant* stage results when the core of the star contracts, heats up, and causes an increased rate of hydrogen burning in the surrounding shell. The shell expands, and the star becomes a red giant. Because of the enormous temperatures involved, the fusion of elements heavier than helium is possible.

Theoreticians are not sure what happens right after the red giant stage. Depending on conditions, a star may pass through a variable stage or an eruptive stage. Regardless, there is no question that it eventually runs out of its nuclear fuel.

4) The *end* stage begins when the nuclear fusion reactions cease. The final disposition of a star is believed to depend upon its mass at the time that the fusion stops:

   a) mass < 1.5 suns: star becomes white dwarf, then black dwarf.
   b) 1.5 suns $<$ mass < 3 suns: star becomes neutron star.
   c) mass > 3 suns: star becomes black hole.

GALAXIES

A *galaxy* is a huge collection of stars occupying a tremendous volume in space.

Types of Galaxies

There are three types of galaxies, named according to the shape.

1) *Irregular* galaxies have no particular shape. They are composed mostly of young stars, called *population I* stars, which contain numerous heavy elements.

2) *Elliptical* galaxies (figure 17-4a) have shapes which range from spherical to ellipsoidal. (An ellipsoid is a solid that results when an ellipse is rotated about its major axis.) They are made up primarily of old stars, called *population II* stars, which formed from pure hydrogen.

3) *Spiral* galaxies have arclike arms that extend outward from a central nucleus (figure 17-4b). *Barred* spiral galaxies are similar (figure 17-4c) except that the arms are attached to short bars that extend out

THE UNIVERSE

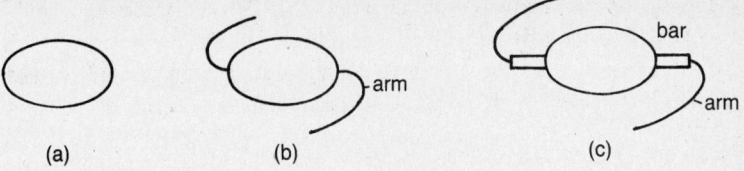

Fig. 17-4 (a) Elliptical galaxy. (b) Spiral galaxy. (c) Barred spiral galaxy.

from the central nucleus. Both types contain population I and population II stars, with the younger stars located in the arms.

There are thousands of *clusters of galaxies* located in the universe. Our solar system is located in a galaxy called the Milky Way (see next section) which is part of a cluster called the *Local Group*.

The Milky Way

The *Milky Way* is a spiral galaxy of about 10^{11} stars, including our sun. A schematic diagram is shown in figure 17-5.

1) The *nucleus* is a mass of stars at the center of the galaxy. It has a diameter of about 10^4 light years.

2) Four or five *arms* spiral out from the center, giving the Milky Way a spread of about 10^5 light years. Stars in the arms revolve about the nucleus, with the outer ones moving more slowly. The sun, located about 2×10^4 light years from the edge of the galaxy, takes about 2×10^8 years to orbit the nucleus.

3) A spherical *halo*, consisting of a thin gas interspersed with globular clusters, surrounds the galaxy. The clusters also revolve around the nucleus, but with motions that are unrelated to the motion of the arms.

Hubble's Law

Hubble's Law relates the velocity at which a galaxy recedes from the earth to its distance from the earth:

$$v = Hd$$

where
v = velocity in km/sec
d = distance in km
$H = 5.5 \times 10^5$ km per sec per parsec
(Hubble's constant)

The astronomer Edwin Hubble, around 1929, made the important discovery that very distant galaxies exhibit red shifts. The more distant the galaxy, the larger the red shift. We know that the magnitude of the red shift increases with velocity. Thus, the more distant the galaxy, the larger the velocity.

A *quasar* is a source of very strong radio wave energy which also has an unusually large red shift. The large red shift implies a large velocity which, by Hubble's Law, means that the object is very far away. This is puzzling; how can an object so far away be so bright? One theory is that quasars are galaxies that are more powerful than ones nearer to us. Another theory is that they are somehow associated with black holes. A handful of astronomers believe that Hubble's Law breaks down for large red shifts and that quasars are really closer than predicted. As yet, there is no convincing evidence for any theory.

Cosmic Radiation

Cosmic radiation ("cosmic rays") is highly energetic radiation that originates in the Milky Way and is incident on the earth's atmosphere. The high energy comes from the acceleration of the cosmic rays as they travel through space and interact with magnetic fields in the galaxy.

1) *Primary cosmic rays* are high speed atomic nuclei as they exist before hitting the earth's atmosphere. Most are hydrogen and helium nuclei, although almost all elements are represented.

2) *Secondary cosmic rays* are particles that result when the primaries collide with the oxygen and nitrogen of the earth's atmosphere. Secondaries consist of neutrons, protons, electrons, gamma rays, and various unstable particles associated with radioactive decay, including mesons. About 75 of these secondaries pass through a square centimeter at sea level each hour.

COSMOLOGY

Cosmology is the study of the origin and evolution of the universe.

The Expanding Universe

The fact that distant galaxies exhibit only red shifts has led astronomers to believe that the universe is expanding. This can be understood with

THE UNIVERSE 207

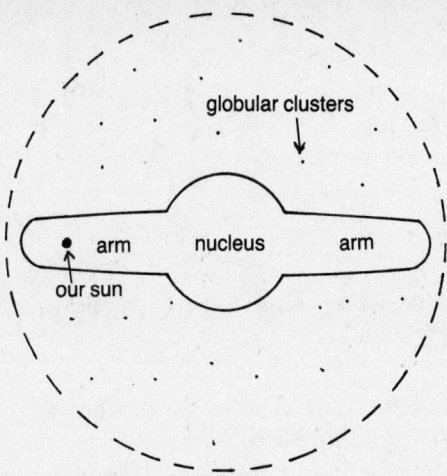

Fig. 17-5 Schematic diagram of the Milky Way

the aid of figure 17-6. The small circle with the five points represents the universe at a particular time. If the universe expands, the points spread out as shown in the larger circle. If you are located at any one of these points, all the other points are receding from you. If this picture were drawn three-dimensionally, one would also see that the more distant points recede faster. Hence, an expanding universe is consistent with the observed facts about distances and velocities of galaxies.

Origin of the Universe

1) The *Big Bang* theory, highly simplified, states that the universe began about ten billion years ago as an extremely hot and dense mass, called a *primeval fireball*. This mass exploded violently and spread out into space. In certain regions, gravitational attraction resulted in the coming together of enough mass to form galaxies. Because the temperature of the original mass was so high, the radiation it emitted must have been in the short-wave region. According to theory, this radiation would today be in the radio wave region due to red shifting. Astronomers have in fact detected this radiation, and their observations agree with the theory. This agreement, plus the fact that the Big Bang theory can account for the expanding universe, makes the theory appealing.

2) The *cyclic* theory states that the universe periodically expands and

Fig. 17-6 Model of expanding universe. Each point gets farther away from all the others as the circle expands.

contracts. The basis of this theory is the assertion that the original explosion was not sufficiently powerful. According to the cyclic theory, gravitational attraction will eventually overtake the forces responsible for expansion and cause the universe to start to contract. All the bodies in the universe will start to converge, lose potential energy, eventually collide, and form a huge mass at a high temperature. This temperature will be great enough to break down all atoms into hydrogen. Eventually, the mass will explode and the whole process will start over again.

Origin of the Solar System

Although no one can say definitely how our solar system started, we can outline a sequence of events that is reasonable in light of known principles and discoveries. Most likely, sometime in the early life of our galaxy a small patch of material condensed, due to gravitational attraction, into a cloud of gas about the size of our solar system. About 90 percent of that gas condensed to form the beginning of the sun. The remaining 10 percent formed a diffuse nebula that eventually cooled, flattened, and broke up into smaller masses called *protoplanets*. These were much larger than present planets because they were mostly gases such as hydrogen and helium. There were some heavier elements, however; they continued to contract toward the center of each protoplanet. Finally, the sun began to shine, having gone through the first steps of the life cycle of a star that we previously described. The heat from the sun boiled away most of the gases and left the relatively dense planetary cores.

PART FOUR

METEOROLOGY

18
The Atmosphere

The *atmosphere* is the envelope of gas that surrounds the earth. A description of the atmosphere at a particular time is called *weather*. Consistent long-term weather patterns are referred to as *climate*. Weather and climate are complex phenomena and are the subject of considerable current research. We shall begin with a look at some of the basic atmospheric variables and then will conclude with a study of how the atmosphere is heated.

COMPOSITION OF THE ATMOSPHERE

The collection of all atmospheric gases, exclusive of water vapor, is called the *dry atmosphere*. We discuss this separately from water vapor.

Dry Atmosphere

The major components of the dry atmosphere are nitrogen (78 percent) and oxygen (21 percent). Except for argon, which occupies about 0.93 percent of the atmosphere, all other constituents are small fractions of a percent. Two of them, carbon dioxide and ozone, are important far out of proportion to their concentration.

1) *Nitrogen* exists in the atmosphere as the diatomic molecule N_2. Although it is relatively inactive, plants can extract it from the atmosphere and make a variety of nitrogen compounds from it.

2) *Oxygen* exists primarily in the diatomic form, O_2. Oxygen is an active molecule; the atmosphere loses it due to reactions with a variety

of substances. However, it gains back oxygen which is given up by various forms of plant life. In the upper reaches of the atmosphere (above 100 km), atomic oxygen, O, is the predominant form. It is produced from O_2 when the molecule absorbs certain wavelengths of ultraviolet radiation and splits into individual atoms. Atomic oxygen is very reactive; in particular, it can combine with O_2 to form O_3, ozone.

3) *Ozone* is a pungent and corrosive gas whose presence in the lower atmosphere is highly undesirable. Fortunately, it is found mainly in the region from 25 km to 50 km. The importance of ozone lies in its ability to absorb certain wavelengths of ultraviolet radiation. This prevents the radiation from reaching the earth where it could do serious harm to living things. Ozone has been the subject of much research lately. This is because pollutants, such as those emitted from supersonic aircraft and spray cans, can reduce the concentration of ozone and hence the absorptive ability of the ozone layer. Just how serious the problem is, remains a question.

4) *Carbon dioxide* not only is important for plant life, but is an absorber of heat as well. Although its concentration is usually listed as constant (.03 percent), its percentage of the atmosphere has been slowly creeping up, due to the burning of fossil fuels.

Moisture

The atmosphere contains moisture in all three states (water vapor, water, and ice). We are primarily interested in water in the vapor state. The air can normally hold up to a certain theoretical maximum of water vapor. This maximum increases with temperature. When it is reached, the air is said to be *saturated*. Below the maximum, the air is *unsaturated*. Sometimes, the theoretical maximum can be exceeded slightly. In that case, the air is *supersaturated*.

1) The *quantity* of vapor can be specified in numerous ways. Two important ones are relative humidity and mixing ratio.

   a) The *relative humidity, RH,* indicates how close the air is to saturation:

$$RH = \frac{\text{grams of vapor in a given volume of air}}{\text{grams of vapor in same volume at saturation}}$$

THE ATMOSPHERE

b) The *mixing ratio* indicates the actual amount of vapor in a given volume of air:

mixing ratio = grams of vapor per kilogram of dry air

2) The *importance* of water vapor can be appreciated by examining the following quantities:

a) *Latent heat* is liberated when water vapor condenses and is absorbed when liquid water evaporates. This amounts to about 2400 joules per gram. This heat is a source of energy for various storms. It also causes air to be more buoyant, an important factor in vertical air motion. Furthermore, evaporation in one area and condensation in another is a means of transporting energy between two locations.

b) *Buoyancy* is directly affected by the presence of vapor; moist air is less dense than dry air. The molecular weight of dry air is about 28.9. This is essentially the weighted average of nitrogen (28) and oxygen (32). The molecular weight of water is only 18. The weighted average for moist air must therefore be lower than for dry air. Because the moist air is more buoyant, the interaction of moist and dry air results in the lifting of the moist air. We will see that this has important consequences in certain storms.

c) *Absorption* of heat is an important function of water vapor. A heavy cloud cover at night, for instance, helps the earth retain some of its heat.

VERTICAL STRUCTURE OF THE ATMOSPHERE

Vertical structure refers to the variation with height of a particular characteristic.

Temperature

Because of distinct warm and cold regions, temperature variation (figure 18-1) is a good way to profile the atmosphere in the vertical. The temperature used is measured at various times throughout the year at many different latitudes and longitudes, then averaged together.

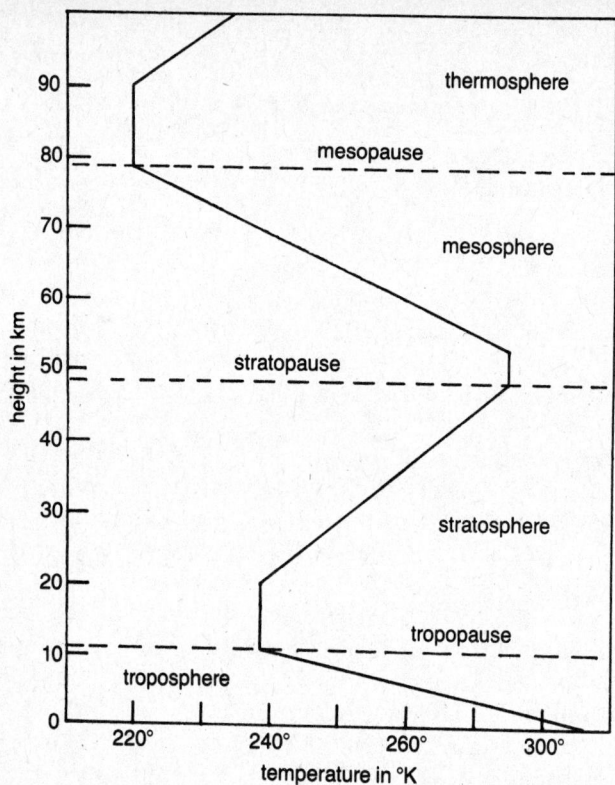

Fig. 18-1 Vertical temperature profile of atmosphere

1) In the *troposphere*, the lowest level, the temperature decreases with height. The top of the layer, called the *tropopause*, is the level at which the temperature stops decreasing with height. The altitude of this level is about 11 km, although at the poles it is lower and at the tropics it is higher. The troposphere is characterized by strong vertical mixing of air masses.

2) The *stratosphere* is the layer just above the troposphere. In the lower part, temperature is approximately constant with height. However, in the upper part, temperature increases with height. This increase comes to a halt at a level called the *stratopause*, about 48 km high. In the stratosphere, very little vertical mixing takes place.

3) The *mesosphere* lies just above the stratosphere and extends to the

THE ATMOSPHERE

mesopause, located at about 80 km. Temperature drops with height in this region; the mesopause is about 40°K colder than the tropopause.

4) The *thermosphere* is the region above the mesopause. It is characterized by a rapid rise of temperature with height. In this region there are layers of ionized gases commonly referred to as the *ionosphere*. The ionization takes place because of absorption of very short wave solar radiation. These layers have the ability to reflect radio waves. Figure 18-2 shows how radio communication can take place over long distances as a result of reflection. The direct wave, *DW*, is attenuated after a relatively short distance. The reflected wave, *RW*, bounces off the ionized layer and ends up far away.

Pressure

Atmospheric pressure drops off rapidly with height, as seen in figure 18-3. This is not surprising in view of the fact that the pressure at any

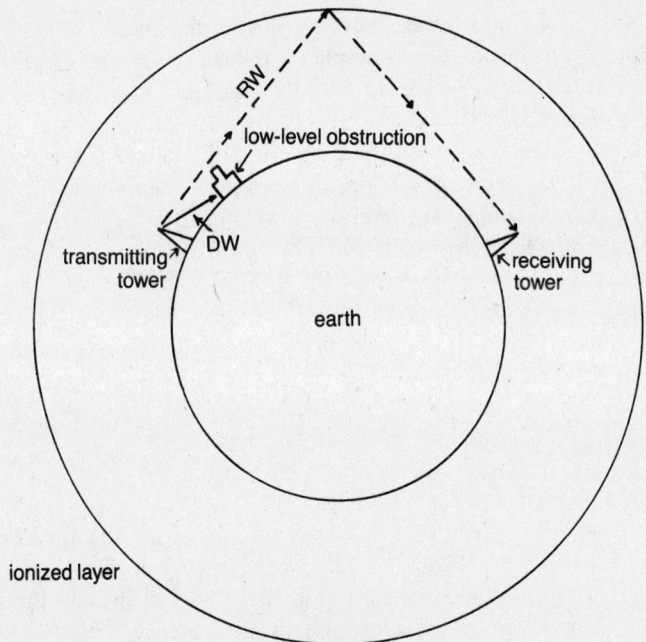

Fig. 18-2 Radiowave reflection off ionosphere

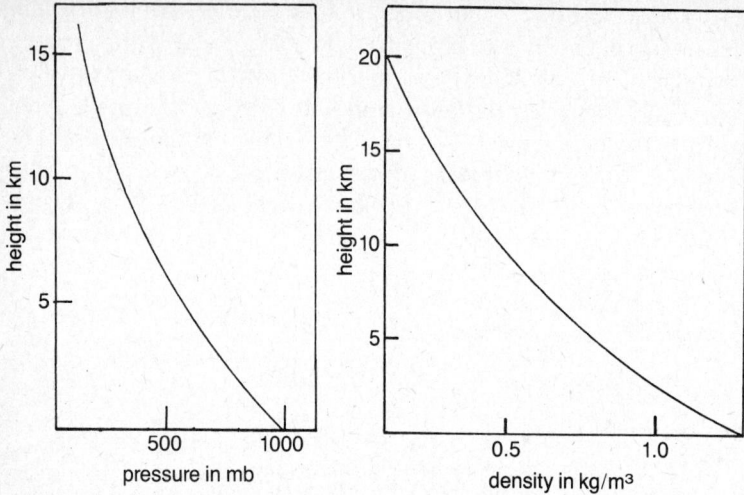

Fig. 18–3 Vertical pressure profile   **Fig. 18–4** Vertical density profile

level is due to the mass of air above it. Most of the mass of air is fairly close to the bottom of the atmosphere, thanks to the pull of gravity and the compressibility of air. In fact, half the mass is confined to the lowest 5.5 km, while 90 percent of it is below 16.5 km.

The standard sea level pressure is about 101.3 kilopascals. A more common unit in meterology is the *millibar*, mb. One kilopascal is ten millibars, so the standard sea level pressure is 1013 mb, often rounded off to 1000 mb. Then the fraction of mass above a given level is just the pressure of that level in mb divided by 1000. For instance, 10 percent of the atmosphere lies above the 100 mb level. (100/1000 = .1.) By the same reasoning, 90 percent of the atmosphere is below the 100 mb level.

Density

Density has a vertical structure similar to pressure, dropping off rapidly with height. This is because of the concentration of mass in the lower part of the atmosphere. Figure 18-4 shows the density profile.

Note that density and pressure cannot be used for dividing the atmosphere into layers. This is because of the absence of any sharp density or pressure changes.

THE ATMOSPHERE

Mixture Uniformity

The various gases of the dry atmosphere are well mixed, up to about 80 km. This means that the relative concentration of a particular gas is the same at any level. Above 80 km, the gases are distributed in a different manner. Those with the higher molecular weights are found at lower altitudes, and vice versa. Thus, it is easier for the lighter constituents to escape.

ATMOSPHERIC HEATING

Sources of Radiation

1) *Solar radiation* is the ultimate source of the energy for all atmospheric processes. The sun can be regarded as a blackbody radiator whose temperature is 6000°K. Most of the energy is in the short wavelength range, with a peak at 0.47 μm (1 μm = 10^{-6} m).

2) *Terrestrial radiation* originates at the earth's surface, which can also be regarded as a blackbody radiator, in this case with a temperature of 250°K. Most terrestrial radiation is in the long wavelength range, with a peak at about 10 μm.

3) *Atmospheric gases* can absorb both long and short wave radiation but only in certain narrow wavelength bands. The radiation that the gases emit, however, is in the long wave region.

Radiation Balance

When incoming radiation exceeds outgoing radiation, the atmosphere heats up. When outgoing radiation exceeds incoming radiation, the atmosphere cools off. For the earth as a whole, the incoming and outgoing radiations, when averaged over a year, are approximately in balance. Hence, there is essentially no change in the average temperature of the earth. There are, however, spatial and temporal imbalances in the incoming and outgoing radiations.

1) There is a *latitudinal imbalance* of radiation which is very important. Below about 37° latitude, the incoming radiation exceeds outgoing radiation. Above 37°, the opposite is true. On the basis of radiation received, we would expect the lower latitudes to get hotter

and hotter, while the higher latitudes get colder and colder. The lower latitudes are indeed warmer than the higher ones, but they are not continually heating up. The excess energy due to radiation must somehow be transported poleward. The mechanism for accomplishing this is an extremely vital one and will be discussed in the next chapter.

2) The *daily temperature lag* is the time difference between maximum incoming radiation and maximum air temperature. Assume the sun rises at 6 A.M. and sets at 6 P.M. Then the maximum incoming radiation is at noon. However, the maximum temperature is not at noon. It is not until about 3 P.M. that the outgoing radiation finally exceeds the incoming radiation for that day. Hence, the atmosphere heats until 3 P.M.

3) The *seasonal temperature lag* is similar. The maximum daily solar radiation occurs about June 21. However, the incoming radiation is greater than the outgoing radiation until about the beginning of August. Hence, the days with the highest average temperatures are generally not until August.

The Radiation Budget

Radiation can be absorbed, reflected, and transmitted. A *radiation budget* is a summary which shows how much of each process occurs so that the incoming and outgoing radiations balance. Let us investigate how the radiation impinging on the earth gets disposed of by the atmosphere and the surface.

1) About 35 percent of the incoming radiation is reflected back to space:

   a) reflection by cloud layers = 24 percent
   b) scattering by the atmosphere = 7 percent
   c) reflection from the ground = 4 percent

2) About 18 percent of the incoming radiation is absorbed by the atmosphere itself:

   a) absorption by the stratosphere = 3 percent
   b) absorption by the troposphere = 13 percent
   c) absorption by clouds = 2 percent

3) About 47 percent of the incoming radiation is transmitted through the atmosphere and absorbed by the ground.

THE ATMOSPHERE

If the surface did nothing but absorb radiation, it would continually heat up. This does not happen. The surface emits the radiation it absorbs so that there is a balance between incoming and outgoing radiation. Much of the radiation emitted by the surface is absorbed by CO_2 and water vapor in the troposphere.

The atmosphere also emits an amount of radiation equal to what it absorbs. Some is emitted directly to space, but some also is directed downward, where it is absorbed by the surface. Thus, the surface receives more energy than it would if there were no atmosphere.

Absorption and the Vertical Temperature Profile

Without getting into the many fine details, we can get a rough idea of the cause of the warm and cold regions of the atmosphere. The key lies in the absorption (or lack of it) in certain layers.

1) The *mesopause is cold* because the thermosphere is an excellent absorber of radiation whose wavelength is less than 0.18 μm. As the incident radiation penetrates deeper into the atmosphere, less of this radiation is available for absorption. By the time the mesopause is reached, very little of any radiation is being absorbed, so the mesopause is cold.

2) The *stratopause is warm* because ozone at 50 km absorbs 0.3 μm radiation very strongly. Even though the maximum concentration of ozone is at 30 km, so much absorption takes place at the top of the ozone layer that the 0.3 μm radiation is weakened considerably by the time the incident radiation gets deeper into the atmosphere. Absorption, and hence heating, diminish below 50 km.

3) The *tropopause is cold* for two reasons. First, almost all the incident radiation that can be absorbed has been absorbed by this level. Second, the troposphere is heated by radiation from the surface that is absorbed by carbon dioxide and water vapor. Most of this radiation is absorbed by the time it gets up to the tropopause. Hence, relatively little absorption of any kind is going on at the tropopause, so it is cold.

19
Atmospheric Motion

In this chapter we shall study air in motion, treating horizontal and vertical motions separately. Air moving horizontally with respect to the earth's surface is called *wind*. This is a vector quantity whose direction is specified by a convention unique to meteorology: winds are identified by the direction from which they come. Thus, an *easterly* wind is one that comes *from the east*.

In discussing air motion, it is convenient to focus on a particular mass of air and follow it in its travels. We call this mass a *parcel* and the rest of the air the *environment*. It is assumed that the parcel neither mixes nor exchanges heat with its environment. In reality, some of each takes place; however, the assumptions are often a good approximation to reality.

HORIZONTAL FORCES

Wind behavior depends on the effects of the forces described below.

Pressure Gradient Force

The *pressure gradient force* is a vector whose magnitude is equal to the pressure difference between two points, divided by their separation. The direction is from the higher to the lower pressure. A given pressure difference can produce a relatively large force if it acts over a small distance; if it acts over a large distance, the resulting force will be relatively small. Pressure gradient forces are the cause of all winds.

ATMOSPHERIC MOTION

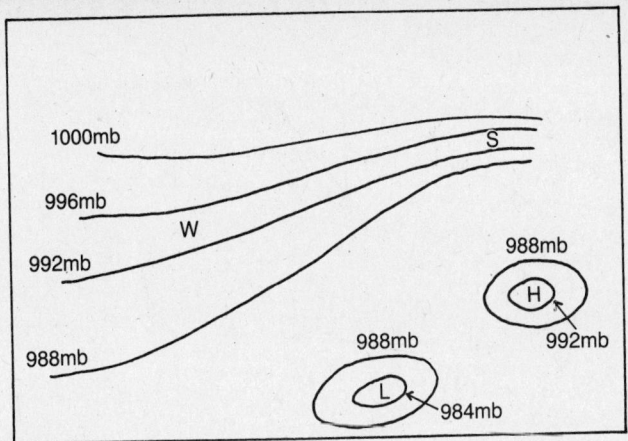

Fig. 19-1 Isobars. View is looking down at surface.

Measurements of pressure over a given region are often pictured as shown in figure 19-1. A line called an *isobar* is drawn so that it connects points of equal pressure. When the isobars are drawn in equal pressure steps, say every 4 mb, the nature of pressure gradients can be noted at a glance. In region S, the pressure difference of 16 mb is exerted over a shorter distance than for region W. Hence, the pressure gradient force for region S is stronger than for region W. In region H, where the isobars are closed, the pressure diminishes as one goes out from the center. This region is called a *high*. Just the opposite happens in the vicinity of L, where the pressure rises as one moves out from the center. This region is called a *low*.

Coriolis Force

All objects moving over long distances above ground in the northern hemisphere appear to veer off to the right. In the southern hemisphere, the veering is to the left. This behavior is called the *Coriolis effect* and arises because of the rotation of the earth.

Any arbitrary area in the northern hemisphere has some counterclockwise rotational motion about the local vertical direction. Figure 19-2 shows the angular velocity vector, ω, for the earth. From the definition of angular velocity, we know that the angular velocity vector

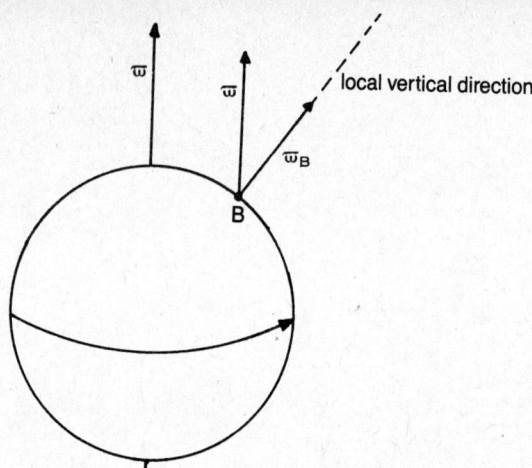

Fig. 19-2 Component of angular velocity, $\overline{\omega}_{B_1}$, at an arbitrary location

is always coincident with the axis of rotation. Thus, it is easy to see that at the north pole, the surrounding area rotates counterclockwise about the local vertical direction. What about other areas?

At point B in figure 19-2, we have drawn a line representing the local vertical direction. By resolution of vectors, we see that some part of the earth's angular velocity is in the local vertical direction. This means that the area centered at B must be rotating around the local vertical. The rate of rotation is slower than once per 24 hours because only a fraction of the angular velocity of the earth is in the local vertical direction. As we move from the pole to the equator, this fraction decreases, becoming zero at the equator. Hence, the Coriolis effect diminishes as one moves from the pole to the equator.

Why does an object in the northern hemisphere appear to veer off to the right? If we throw an object horizontally in a straight line, it continues in a straight line because there are no horizontal forces acting on it. (Gravity makes it fall, but we are only interested in what happens horizontally.) We on the earth turn counterclockwise. Hence, we really veer off to the left of the moving object. Because we view all motion from the earth's frame of reference, we think that the object veers to our right. We conclude that a force causes this motion and call it the

ATMOSPHERIC MOTION

Coriolis force. There really is no force; because we are in a rotating coordinate system, we must imagine that a force exists so that we can use Newton's laws to analyze the motion.

The Coriolis force increases as the speed of the air parcel increases. However, the practical effects are noticeable only if the parcel travels great distances.

Centrifugal Force

Air parcels do not necessarily move in straight lines. Those that move in circular paths have a centrifugal force associated with them. The centrifugal force on these air parcels arises for reasons which we have discussed previously.

Friction

Friction affects air parcels the same way as other objects: it slows them down. In the atmosphere, friction effects are particularly important for the lowest kilometer of air. This is because the topography of the surface interferes with motion in this layer.

CLASSIFICATION OF WINDS

Types of Horizontal Motion

Certain kinds of horizontal motion are given special names in meteorology.

1) *Convergence* results when air particles are forced closer together.
2) *Divergence* is the spreading apart of air particles.
3) *Cyclonic* flow is motion that is curved in the same sense as the earth's angular velocity. In the northern hemisphere this is the counterclockwise direction.
4) *Anticyclonic* flow is curved in a sense opposite to the earth's angular velocity. In the northern hemisphere this is the clockwise direction.

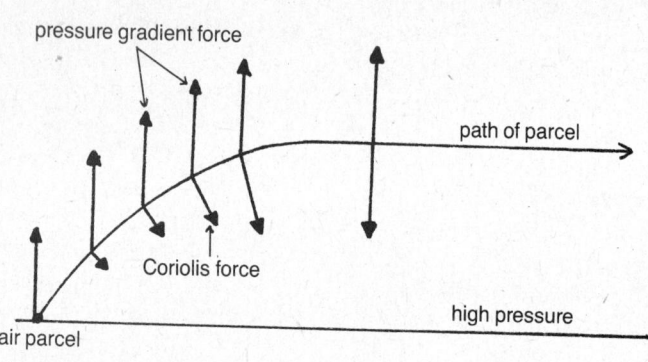

Fig. 19-3 Approach to geostrophic wind. Note that the Coriolis force is always to the right of the motion and increases as the parcel picks up speed.

Frictionless Geostrophic Wind

A *geostrophic wind* is one where the Coriolis and pressure gradient forces balance. The centrifugal force can be neglected because the motion is essentially in a straight line. Figure 19-3 shows how a geostrophic wind develops. A parcel of air located initially in a region of high pressure starts to move directly toward the low pressure isobar. As it picks up speed, the Coriolis force increases, deflecting it to its right. This constant deflection has the effect of changing the direction of the Coriolis force until it opposes the pressure gradient force. By this time, the parcel has picked up enough speed so that the magnitude of the Coriolis is exactly the same as that of the pressure gradient force. The parcel simply continues parallel to the isobars.

Frictionless Cyclostrophic Wind

In a *cyclostrophic wind,* there is a balance between pressure gradient and centrifugal forces. The Coriolis force can be neglected because the horizontal extent of this motion is not large. A tornado, which we shall study later, is an example of a cyclostrophic wind.

Figure 19-4 shows how both clockwise and counterclockwise motions are possible. The centrifugal force (pointing outward) and the pressure gradient force (pointing inward) are in balance in each case.

ATMOSPHERIC MOTION

Frictionless Gradient Wind

In the *gradient wind*, there is a balance of pressure gradient, Coriolis, and centrifugal forces. The pressure gradient points toward the low, the centrifugal force is outward, and the Coriolis force is to the right of the motion. Figures 19-5a and 19-5b show how gradient winds are possible around both high and low pressure regions. Around a low, the motion is cyclonic. This is because an air parcel, initially moving across the isobars to low pressure, gets deflected to its right by the Coriolis force. This causes it ultimately to move in the counterclockwise direction, parallel to the isobars. Around a high, the motion is anticyclonic. A parcel of air initially moving from high to low across the isobars gets deflected to its right, which, in this case, causes it to move clockwise.

The Effect of Friction

We have seen how frictionless winds follow the isobars. When friction is taken into account, the flow is slightly cross-isobar, going toward the low pressure. We can see why this is so by looking again at figure 19-3. The speed of the wind is such that the Coriolis force exactly balances the pressure gradient force. Friction, of course, slows the wind somewhat. This, in turn, reduces the Coriolis force. Since the pressure gradient force is still the same, there is a net force on the parcel toward the low pressure isobar.

The same reasoning holds for the gradient case, with cross-isobar flow introducing a radial component. Friction in a gradient wind causes air flowing around a low to spiral towards the center (figure 19-5c). Friction in a gradient wind causes air flowing around a high to spiral out from the center (figure 19-5d).

Having reviewed all the forces that affect motion, we should note that only the Coriolis effect is hemisphere-dependent. Since we live in the northern hemisphere, our descriptions of motion involving the Coriolis effect will be for that hemisphere, unless otherwise stated.

THERMALLY DRIVEN CIRCULATION

A *thermally driven circulation* is one which is caused by horizontal differences in temperature. Figure 19-6 shows a cross section of a hypothetical area where there is a uniform temperature on a horizontal

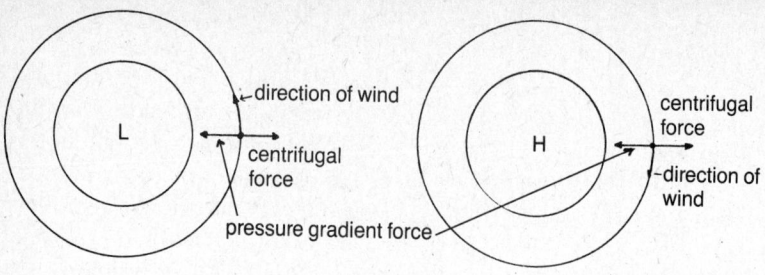

Fig. 19–4 Cyclostrophic winds. Note that clockwise and counterclockwise motions are consistent with the balance of forces.

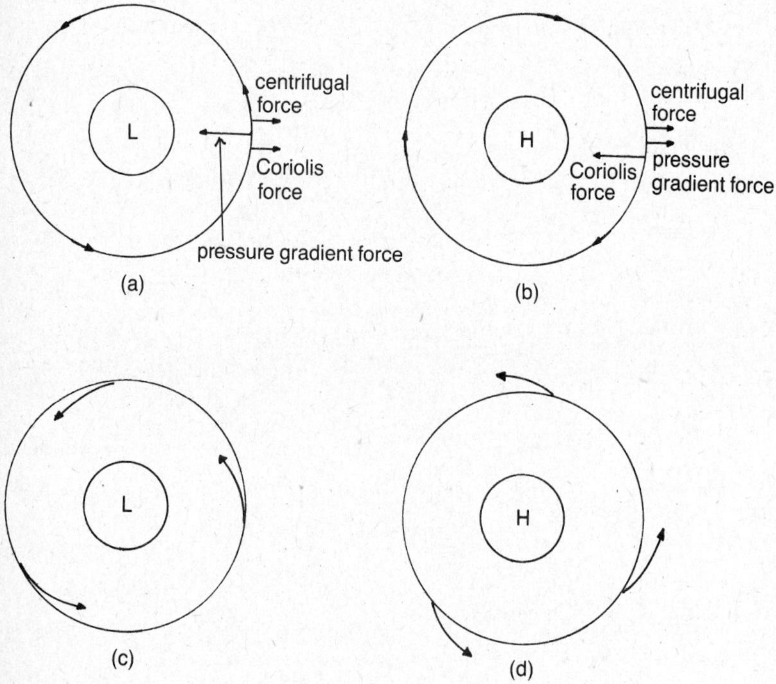

Fig. 19–5 Gradient winds around highs and lows. Balance of forces is shown in (a) and (b); effect of friction on motion is shown in (c) and (d).

ATMOSPHERIC MOTION

surface. The isobars are parallel to the surface because the pressure decrease with height is the same over all parts of the surface. In figure 19-7, we see the same surface with one side heated and the other side cooled. The column of air over the heated side rises because the heated air expands. The column of air over the cooled side sinks because cooled air contracts. The result is that the isobars are no longer parallel to the surface. We can understand this by remembering that the pressure at any level depends on the weight of the air above. Let us consider the level above which we find half the column's weight. Call the pressure here P_0. When the column expands, this level must rise in order that half the weight remain above it. Hence, the isobar passing through P_0 must go up a bit. By similar reasoning, the isobar passing through P_0 in the cool region must dip down.

Now we can see how a circulation develops. At a given height above the surface, the pressure in the warm column is higher than in the cool one. Air at the high levels then flows from the warm side to the cool side. This increases the surface pressure on the cool side because the weight of the column of air on that side is increased. At the surface, the higher pressure on the cool side forces the air from the cool to the warm side. The complete circulation is shown by the arrows in figure 19-7; the horizontal motions just described are complemented by rising warm air and sinking cool air. Since warm air is brought to the cool region, and vice versa, the effect of the circulation is to even out the temperature differences that caused it in the first place.

Two things should be noted. First, the sharper the temperature contrast, the stronger the pressure difference and resulting wind. Second,

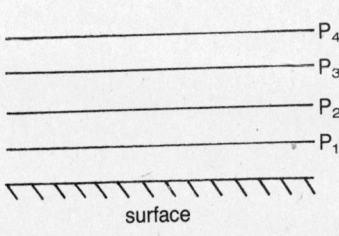

Fig. 19–6 Isobars on a surface with uniform temperature. $P_4 < P_3 < P_2 < P_1$.

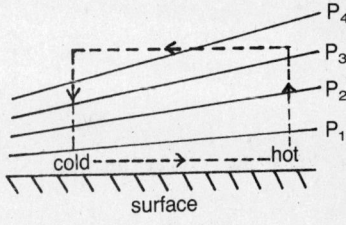

Fig. 19–7 Isobars on a surface with nonuniform temperature. $P_4 < P_3 < P_2 < P_1$.

the pressure difference between warm and cold regions at a given level increases with height. Hence, the wind speed increases with height.

We shall see that the thermally driven circulation has application on a global scale, although serious modifications must be made due to the rotation of the earth. A simple application, however, is the *sea breeze*. The sea heats up much more slowly than the land. During the day, when the sun bears down on both the sea and the land, the land gets hotter. Thus, we have a situation like the one shown in figure 19-7, with the sea as the cool side. From our previous discussion, it is clear that the wind must blow from sea to land. This is why it is usually cooler in summer at the shore than inland. At night, the land cools off much faster, so the breeze direction is reversed. Cool air flows out to sea.

GENERAL CIRCULATION

If we observe the winds in a particular region, we note that they change from day to day. However, if we average our observations over a long period of time and over large areas, the large-scale, long-term motions stand out. We refer to these motions as the *general circulation*. While the general circulation is quite involved, we will concentrate on two main questions:

1) What do the wind patterns look like at the surface and in the upper air?
2) How do these winds help maintain the global heat balance?

In discussing the general circulation, it is convenient to resolve the horizontal motion into two perpendicular components. *Zonal* flow is motion parallel to circles of latitude. *Meridional* flow is parallel to circles of longitude.

Zonal Wind Belts at the Surface

There are three pairs of wind belts, separated by low and high pressure regions, that are obtained by the kind of averaging just described (figure 19-8).

1) The *trade winds*, also known as the tropical easterlies, extend from the equator to about 30°.

ATMOSPHERIC MOTION

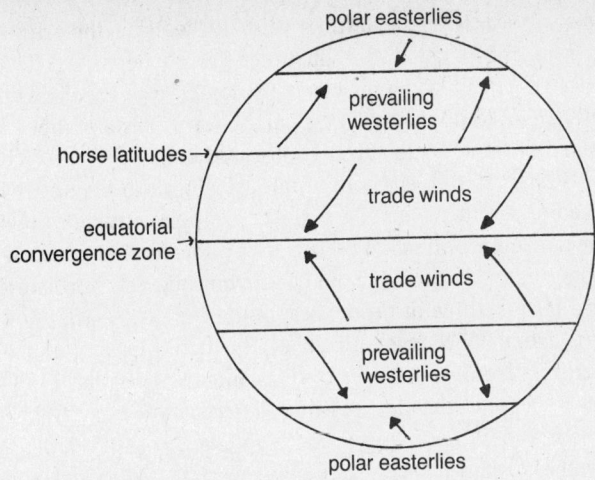

Fig. 19-8 Global wind belts

2) Just poleward of the trade winds are the *horse latitudes*. These are regions where the winds are almost nonexistent. (In the early days of commerce, sailing ships would get stranded here and throw their cargo of horses overboard to save food and water.)

3) Between the belts of trade winds is the *equatorial convergence zone*, a region of low pressure and calm winds.

4) In the mid-latitudes, from about 30° to 60°, the *prevailing westerlies* blow. The mid-latitudes are characterized by traveling high and low pressure areas. These features do not show up in the general circulation because they are masked by the averaging process. We will make a separate study of these motions in the next chapter.

5) At the polar regions, the winds are easterly and are called the *polar easterlies*.

6) Between the polar easterlies and the prevailing westerlies is another zone of convergence called the *polar front*.

Upper Air Motion

The motion of the upper tropospheric air is predominantly westerly. There are two main features which are of primary interest: upper air waves and the jet stream.

1) *Upper air waves* is the term we use to describe the path of the upper air westerlies. Figure 19-9 illustrates this. Regions of anticyclonic flow are called *ridges*, while regions of cyclonic flow are called *troughs*. The length of these upper air waves varies: long ones are about 5000-8000 km and short ones are about 1500-3000 km.

2) The *polar jet stream* is a core of fast-moving air in the upper waves. It is located above the polar front. Since the polar front is a region of sharp temperature contrast, it is not surprising to find a strong wind aloft. From winter to summer, the polar front migrates northward, due to the changing latitude of the vertical ray of the sun. The jet does likewise, going from 35°N to about 50°N. Furthermore, the jet's average speed decreases from 130 km/hr in the winter to half that in the summer. This is because the pole-to-equator temperature contrast diminishes in the summer.

The jet stream and upper waves undergo a cycle in which the wave patterns become increasingly undulating. They change from gently curved long waves to sharply curved short ones. Air in the long waves has a strong zonal component, a situation referred to as *high index circulation*. Air in the short waves has a weaker zonal component, and a stronger meridional one; we call this *low index circulation*.

Maintaining the Heat Balance

We have seen that there is a deficit of radiant energy at the poles and an excess at the equator. How does the atmosphere compensate for this imbalance?

1) In the *tropical region*, the air motion is a thermally driven circulation called a *Hadley cell*. This cell works like the one shown in figure 19-7; consider the warm area to be the equator and the cool area to be 30°N. In that case, we see that the upper air moves northward and the surface air moves southward. However, the Coriolis force comes into play and deflects everything to the right of the motion. Hence, the southward moving surface air develops an easterly component, while the northward moving upper air develops a westerly component. Because the Coriolis force is not too strong in the tropics, the deflection to the right is weak. The meridional component is thus more significant than the zonal one. It is the meridional component that transports warm air northward and cool air southward.

ATMOSPHERIC MOTION

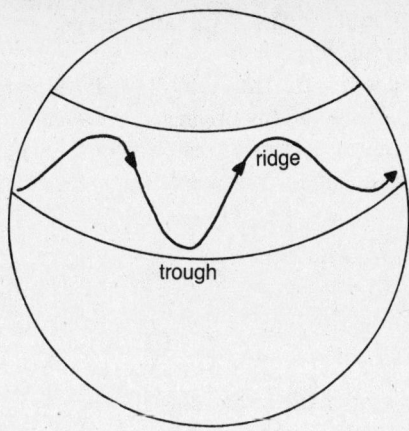

Fig. 19-9   Upper air flow

Why doesn't the cell extend all the way to the pole? By the time the air gets to 30° latitude its westerly component is as fast as friction will allow. Also, on its trip north, it loses about 1°C to 2°C per day. It is cool enough to start sinking at 30° latitude. Hence, the cell is confined to the region indicated.

2) *Outside the tropics,* the Coriolis force is strong. The flow has only a weak meridional component, so the direct flow just discussed cannot be responsible for heat transfer. In this region, turbulent exchanges of large masses of cool and warm air transport heat northward. These exchanges are accomplished by the moving high and low pressure areas prevalent in the mid-latitudes. Also, the upper air waves are involved in heat transport. In figure 19-9, we see that west of the trough, cold air is being pushed southward, while east of the trough warm air is moving northward.

3) In the *polar region,* cold air is dense and tends to subside. When it does, it diverges at the pole and gets pushed into a southeasterly flow by the Coriolis force. Hence, cold air is transported to a warmer region.

By way of summary, we can say that the atmosphere functions as a heat engine, taking in heat at a high temperature and giving it up at a lower one. The ultimate source that drives this engine is the uneven heating of the sun. We have seen how the air rises over a warm region. This elevation of a large mass of air tends to increase the potential energy of the atmosphere. The extra potential energy is converted into the kinetic energy of wind motion. The winds transport energy from the warm to the cool regions; this tends to reduce the temperature contrasts that started the motions originally. Also, energy ultimately dissipates through friction. The only reason why the whole system keeps going is that solar radiation continues to maintain the pole-to-equator temperature contrast.

VERTICAL MOTION

Vertical motion is the result of an imbalance between the gravitational force on an air parcel and the buoyant force. Imbalances can develop because of density changes due to expansion and contraction.

Stability

Air is classified as stable, neutral, or unstable depending on how a parcel of air, originally in equilibrium, reacts to a vertical displacement.

1) The atmosphere is *stable* if the parcel returns to its equilibrium position.

2) The atmosphere is *unstable* if the parcel continues moving in the direction of the displacement.

3) The atmosphere is *neutral* if the parcel simply remains in the displaced position.

Lapse Rate

A *lapse rate* tells how rapidly air cools as its altitude increases.

1) The *environmental lapse rate* refers to the rate at which the troposphere cools with height. This rate varies with local conditions; an approximate average figure is 6.5°C per kilometer.

2) The *dry adiabatic lapse rate* is the rate at which an unsaturated air parcel cools as it gains altitude. This is about 10°C per kilometer. When a parcel of air rises, it expands because the surrounding pressure decreases with altitude. According to the laws of thermodynamics, ex-

panding air cools. Hence, a rising parcel of air cools, even though no heat is exchanged with the environment. Similarly, it can be shown that sinking air parcels warm up at the same rate.

3) The *saturated adiabatic lapse rate* refers to the rate at which a saturated air parcel cools as it rises. This rate varies between 4°C and 10°C per kilometer. In a rising saturated air parcel, excess water vapor condenses because air can hold less and less vapor as it cools. The condensing vapor liberates latent heat; the liberated heat tends to slow down the cooling rate.

When a saturated parcel sinks, it warms up. The amount of water vapor it can hold immediately increases. Hence, a sinking air parcel is always unsaturated and warms at the unsaturated lapse rate.

Environmental versus Adiabatic Lapse Rates

The stability of the atmosphere can be ascertained by comparing the local environmental lapse rate with the adiabatic lapse rate.

1) The atmosphere is said to be *absolutely stable* if the environmental lapse rate is less than both adiabatic lapse rates. This can be explained very simply. If a parcel of air under these conditions is displaced upward, it will cool faster than the environment. Hence, it will be denser than its surroundings and sink back to its original position. If the parcel is displaced downward, it will warm faster than its surroundings and rise to its original position. This behavior is exactly what we have previously defined as stable.

2) The atmosphere is said to be *absolutely unstable* if the environmental lapse rate is greater than the adiabatic rates. In this case, a parcel of air displaced upward does *not* cool as rapidly as its environment. Thus, it is always less dense than its surroundings and continues to rise. If the parcel is displaced downward, it warms more slowly than the environment. Hence, it is always denser than the surroundings and continues to sink. This is what we have previously defined as unstable.

3) The atmosphere is said to be *conditionally stable* if the environmental lapse rate is between the dry and saturated adiabatic rates. If the air is unsaturated, the environmental rate is smaller than the parcel rate; this fulfills the condition for stability. If the air is saturated, the environmental rate is greater than the parcel rate; this fulfills the condition for instability.

20
Weather Disturbances

In this chapter, we will investigate a number of familiar weather phenomena, using some of the physical principles already introduced.

CLOUDS AND PRECIPITATION

A *cloud* is a visible mass of water or ice particles located above the earth's surface. *Precipitation* is the general name given to the variety of ice or water particles that fall to the earth's surface.

Formation of Clouds

Two criteria must be satisfied before clouds can form. First, air must be cooled to the point of saturation. Second, small particles called *condensation nuclei* must be present so that the vapor has a surface on which to condense.

1) Cooling of the air to the saturation point is accomplished by lifting. *Mechanical lifting* occurs when a mass of air flows over a mountain. *Dynamic lifting* occurs when air converges about a low pressure area on the earth's surface. This converging air has nowhere to go but up. *Convection currents*, in which heated air masses rise, can also accomplish lifting.

2) *Condensation nuclei*, such as dust and salt, are abundant in the atmosphere. Salt is particularly effective as a condensation nucleus because it is *hygroscopic:* it has an affinity for water. *Freezing nuclei*, about which ice particles form, are in shorter supply. Often, they are supplied artificially, a process known as *cloud seeding*.

Growth of Cloud Droplets

Cloud droplets grow because condensation on tiny droplets produces larger droplets. Condensation takes place on a flat surface of pure water when the relative humidity is 100 percent. But droplets are obviously not flat; also, they are not pure, because substances are dissolved in them. These two complications have opposite effects on the tendency of vapor to condense onto the droplets.

1) The *solute effect* refers to the enhanced ability of vapor to condense on a droplet which has something dissolved in it. The presence of solute lowers the vapor pressure of the droplet. This is because solute molecules take the place of water molecules at the droplet's surface. Fewer water molecules are there to escape so the vapor pressure is lower. The lowered vapor pressure makes it easier for an approaching vapor molecule to condense on the droplet. It should be noted that if the solute effect were the only one present, condensation could take place at less than 100 percent humidity.

2) The *curvature effect* refers to the tendency of sharply curved surfaces, such as very small droplets, to inhibit condensation. The vapor pressure of a small droplet is rather high when compared to that of a flat surface. This is because a given molecule at the surface of the drop has fewer neighboring molecules to attract it, so it has a greater likelihood of escaping. The increased vapor pressure makes it more difficult for an approaching vapor molecule to condense on the droplet. Note that if the curvature effect were the only one present, the humidity would have to be greater than 100 percent before condensation could take place.

Consider a tiny droplet, whose initial solute concentration is great. The solute effect is then so strong that it overcomes the curvature effect and permits condensation to take place at about 97 percent humidity. With the growth of the droplet, the concentration decreases and the radius increases, weakening both effects. However, the solute effect weakens more than the curvature effect. This means that in the initial stages of growth, condensation becomes more difficult. The relative humidity requirement for condensation then increases, ultimately becoming slightly more than 100 percent. Soon after this figure is reached, the drop has grown large enough (10^{-3} mm) so that neither the solute nor the curvature effect is significant. Then the humidity requirement drops to the normal 100 percent; since the existing humidity is more than 100 percent, the drop simply keeps growing.

Rain

The condensation process is very slow. It would take days for a raindrop, which has a volume of about 10^6 cloud droplets, to form. The actual mechanism for raindrop formation depends on whether the cloud is warm or cold.

1) *Coalescence* is the coming together of many cloud droplets. It is an observed fact that droplets of various sizes exist in a typical cloud. These droplets fall at speeds which increase with size. Sometimes a large droplet runs into a smaller one; at other times, a small one is captured in the wake of a larger one. In either case, many drops coalesce and form a large raindrop.

Coalescence is the process responsible for the formation of raindrops in warm clouds. This is how most rain which originates in the tropics forms.

2) The *Bergeron-Findeisen* process involves condensation on ice particles. In cold clouds, tiny ice crystals coexist with liquid water drops. Because the molecules of ice are held together more tightly than those of water, the vapor pressure of ice is lower. It is therefore easier for molecules of vapor to condense on ice (where they freeze) than on water. Often, a humidity value which is too low for condensation on water is more than enough for condensation on ice. Hence, the ice particles grow. The ice then falls, melting as it goes through a warm layer of the atmosphere.

Most of the rain that falls in the mid-latitudes forms in the above manner. This is because the clouds extend high enough to be in subfreezing temperatures.

Other Precipitation

1) *Snow* develops from water vapor that solidifies without going through the liquid state. This produces tiny, six-sided ice crystals whose growth is particularly fast at each apex. This is why mature snowflakes have hexagonal forms.

2) *Sleet* is rain that falls through a very cold layer of air and freezes. It should not be confused with *freezing rain*, which is rain that freezes when it hits the ground.

3) *Hail* consists of ice particles that fall and rise due to the updrafts and downdrafts that are found in thunderstorms. The up and down

WEATHER DISTURBANCES

cycles enable the ice particle to take on layer after layer of moisture, each of which freezes as it is taken on.

4) *Fog* is essentially a cloud located at the earth's surface. It occurs any time the air cools to the saturation point. *Dew* is water that condenses at the earth's surface when the temperature drops low enough for the air to saturate. This temperature is called the *dew point*. *Frost* is ice that forms at the earth's surface without passing through the liquid stage. It is *not* frozen dew.

THE MID-LATITUDE CYCLONE

A *cyclone* is a moving low pressure region. It is the most common weather disturbance in the middle latitudes. In this section, we will investigate its structure, development, and relation to the upper air waves.

Air Masses and Fronts

1) An *air mass* is a large volume of air that has fairly uniform temperature and humidity characteristics. These characteristics develop when the air is situated for some time over a particular region, called a *source region*.

2) A *front* is the boundary separating two air masses. Although the transition from one air mass to another is not discontinuous, the frontal region is rather sharply defined.

In a *cold front,* there is cold air advancing toward warm air. Since the cold air is denser, it pushes the warm air up. A cold front is symbolized by a line with triangles pointing in the direction of air mass motion:

Cold fronts usually produce heavy, violent precipitation because the warm air is pushed upward vigorously.

In a *warm front,* it is the warm air that advances toward the cold. The warm air glides gently over the layer of cold air. A warm front is symbolized by a line with semicircles pointing in the direction of motion:

Warm fronts usually have light, steady precipitation because the warm air is pushed up slowly.

In a *stationary front*, neither air mass approaches the other. The symbol for this front is a line with alternating triangles and semicircles:

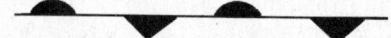

Figure 20-1 shows cross sections of warm and cold fronts. The slopes of the fronts are exaggerated in the drawing; however, the figures given in parentheses are typical values.

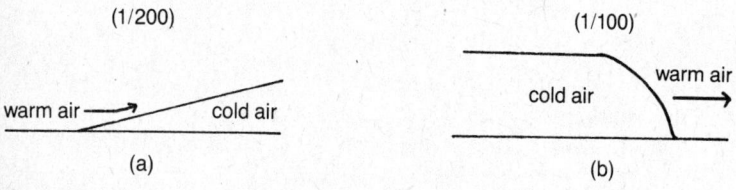

Fig. 20–1 Warm (*a*) and cold (*b*) fronts. Numbers in parentheses are slopes of typical fronts.

Stages of a Mid-Latitude Cyclone

Figures 20-2 through 20-6 show various stages of a typical mid-latitude cyclone, from beginning to end. These are described in order below.

1) Figure 20-2: The cyclone is born in the region of the polar front, which separates cold easterlies from warm westerlies. A small perturbation, which will grow in amplitude, results in a tongue of warm air intruding into the cold.

2) Figure 20-3: The amplitude of the perturbation grows. The pressure over it drops, resulting in a low pressure center with the appropriate gradient flow (arrows). From the direction of the flow, we see that on the west side of the perturbation there is a cold front, while on the east side there is a warm front. As the perturbation and low pressure center both intensify, the cold front, which moves faster, starts to close in on the warm front. A cross section through line KK', seen in figure 20-4, shows both fronts with their attendant precipitation. Note that between them is a clear area of warm air.

3) Figure 20-5: At this point, the cold front has caught up with the warm front to produce an *occluded front*. This is seen in a cross section through line MM', figure 20-6, which shows the warm air lifted entirely

WEATHER DISTURBANCES

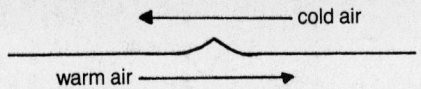

Fig. 20–2 Top view of perturbation at polar front

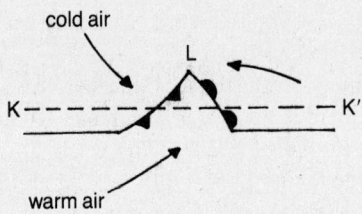

Fig. 20–3 Top view of the intensification of a mid-latitude cyclone

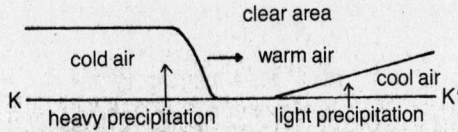

Fig. 20–4 Cross section through line KK' of fig. 20–3

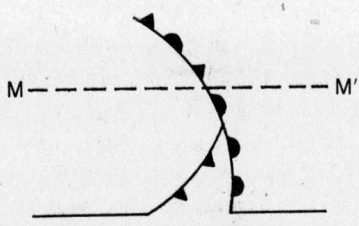

Fig. 20–5 Top view of occluded stage, shown with symbols for occluded front

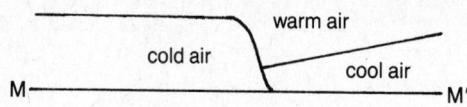

Fig. 20–6 Cross section of line *MM'* of fig. 20–5

off the ground. From the cross section, we can see that what was a clear area is now filled with precipitation.

This is the beginning of the end of the cyclone. With no further lifting of moist air, there is no longer a source of latent heat to provide the cyclone with a source of energy. The small lip of warm air left protruding into the cold simply dies out, leaving an undisturbed front between the two air masses.

Cyclogenesis

Cyclogenesis is the creation of a new cyclone. Although the details of this process are quite complicated, the broader aspects are not difficult to understand.

Cyclones are intimately related to the upper air waves. We have noted that the wavelength of these waves can vary between long (high index circulation) and short (low index circulation). According to the laws of fluid dynamics, shortening of the wavelength results in the development of regions of divergence and convergence along the path of the jet stream. The path of the wave just to the west of the trough is a region of convergence; just to the east is a region of divergence. The consequences of this are illustrated in figure 20-7, which is a cross section of the atmosphere showing surface and upper air motions. Air that diverges at the upper level is replaced by rising air from the lower level. Similarly, air that converges at the upper level sinks and diverges at the surface. The region of divergence at the surface is a high pressure area with sinking air rotating anticyclonically. Since sinking air is unsaturated, the surface high is a region of clear skies. The region of convergence at the surface is a low pressure area with rising air rotating cyclonically. The rising air has the opportunity to become saturated; this region is often cloudy, with precipitation. The cyclonic rotation

WEATHER DISTURBANCES

results in winds that are just like those pictured in figures 20-2 through 20-6. Hence, the region at the surface which is underneath the upper level divergence is an ideal location for a mid-latitude cyclone.

Now we can see that cyclogenesis is the result of a close relationship between upper air and surface events. The very sharp temperature contrast that exists in the region of the polar front is responsible for the jet stream that lies above it. The jet stream is in turn responsible for the development of moving low pressure areas in the region of the polar front.

Severe Weather

1) *Thunderstorms* are storms that develop from *cumulonimbus clouds*. These are clouds that result when a deep layer of moist air is lifted to great heights. The tops of these clouds are at subfreezing tem-

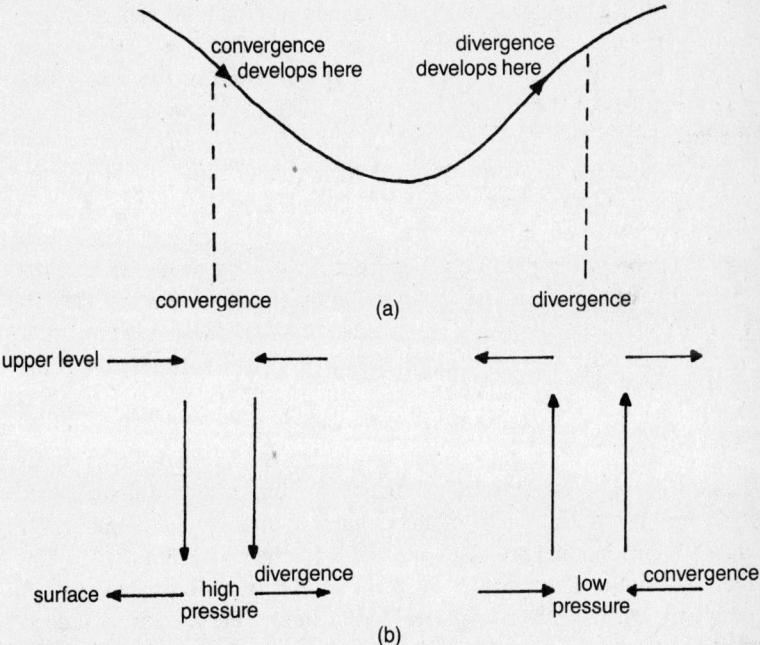

Fig. 20–7 (*a*) Top view of jet stream. (*b*) Cross section of atmosphere showing motions below jet stream.

peratures. Thunderstorms occur when the atmospheric conditions are warm, humid, and unstable. In addition to rain, these storms produce thunder and lightning. Thunderstorms last only one or two hours; this short lifespan can be divided up into three stages:

a) In the *developing stage*, a large mass of air rises rapidly to heights where the temperature is below freezing. Water droplets and ice particles build up in the cloud. Eventually they fall because the rising air cannot hold them up. A heavy rain commences, dragging air along with it to create downdrafts.

b) In the *mature stage*, updrafts and downdrafts exist simultaneously. The precipitation cools the downdrafts so that they spread out when they hit the ground. This is why it is so gusty during a thunderstorm.

c) The *dissipating stage* sets in when the downdrafts overwhelm the updrafts. The rain subsides, and the cloud breaks up.

Very often, thunderstorms exist as several clouds of the type just described. Each cloud is referred to as a *cell*. The downdraft, which hits the ground, often pushes up air that is in front of the cloud. This upward moving air can form a new cell, thereby extending the life of the thunderstorm.

Lightning is an electrical discharge between a thundercloud and the ground. A complex series of events results in a separation of charges so that the lower part of the cloud becomes negatively charged. This negative charge repels negative charges from the ground below. Hence, the ground becomes positively charged. A tremendous electric field is created; when the strength of the field exceeds the insulating capability of the air, a discharge occurs. This heats up the air in the region of the discharge so suddenly that a shock wave is produced. The result is called *thunder*.

2) A *tornado* is a violent storm that consists of a narrow column of rapidly spinning air extending downward from a large thundercloud. The pressure at the center of the column is often about 25 mb lower than the surrounding pressure. The winds can be several hundred kilometers per hour. The pressure drop makes the tornado particularly destructive. When a tornado passes over a house, the drop is so sudden that the pressure inside the house has no time to drop accordingly. For a short time, then, the force pushing out on the walls and roof is greater than the force pushing in. The house literally explodes.

Not all thunderstorms produce tornadoes. It appears that the updrafts of thunderstorms have to be combined with horizontal crosswinds so that a cyclonic circulation can develop. The area of the Great Plains of the United States is particularly favorable for these conditions. In the west, the Rocky Mountains prevent surface air from reaching the plains. Only the relatively cool and dry air passing over the mountains makes it into the area. In the east, over the Atlantic at about 30° latitude, a persistent high pressure area causes air to flow over the Gulf of Mexico and onto the plains. Hence a cool, dry westerly flow meets a warm, moist southerly flow. Not only is the situation unstable from the standpoint of vertical motion, but the necessary horizontal crosswinds are also present.

3) A *hurricane* is a tropical cyclone. In the tropics, small waves called *easterly waves* are often superimposed upon the tropical easterlies. *Shear lines*, which are regions of sharp wind shift, are also prevalent. It is not known why, but some of these disturbances turn into centers of low pressure that intensify and turn into hurricanes.

A full-blown hurricane is about 500 km in diameter. It has an area in the center about 30 km in diameter whose pressure is very low. This is called the *eye*. Surrounding the eye is the *eye wall*, a swirling mass of air with tremendous updrafts. These updrafts are so strong that air moving horizontally toward the eye gets swept upward in the eye wall. The only way for air to get to the eye is by subsidence of air above the surface in the eye.

The sole driving force for the hurricane comes from the surface air. The condensing vapor from this air releases a tremendous amount of latent heat energy. This is in contrast to the tornado, whose driving force is located in the thundercloud above the storm, and the mid-latitude cyclone, which has a connection to the upper air via the jet stream.

There are a number of other characteristics that set hurricanes apart from mid-latitude cyclones.

a) Hurricanes form only over water that has a temperature greater than 26.5°C. At lower temperatures, the air cannot hold enough moisture to release the necessary latent heat.

b) Hurricanes originate between the latitudes 5° and 20°. Outside of 20°, the water temperature is always too low. Inside of 5°, the Coriolis force is so weak that disturbances cannot be given a strong enough deflection to impart a significant spinning motion.

c) There are no fronts in a hurricane. This is because there is only one type of air mass present: warm and humid.

d) The eye is warmer than its surroundings and often free of cloudiness. This is due to the subsiding air, which warms as it sinks and is always unsaturated.
e) Hurricanes usually die out when they travel over land. Since water vapor is the source of energy for the hurricane, it is clear that the air over land is too dry to provide much vapor.

PART FIVE
GEOLOGY

21
Introduction to Geology

In this chapter, we begin our study of the solid earth. Although it is sufficient for our purposes to regard the earth as a sphere, we should note that it is actually an oblate spheroid. This is like a sphere that bulges out at the equator and is flattened at the poles. This bulge comes from the rotational centrifugal force, which is a maximum at the equator and zero at the poles.

If we could slice the earth in half, we would see three concentric sections: a central *core,* an intermediate *mantle,* and an outer *crust.* Subsequent chapters will cover the core and mantle. In this introduction, we will be concerned mainly with features of the crust.

MINERALS, ROCKS, AND SOILS

Minerals

A *mineral* is an element or an inorganic compound that occurs naturally on the earth. Most minerals are compounds; elemental minerals like gold or copper are relatively rare. The *silicate minerals* are the most abundant. They are minerals that contain silicon, oxygen, and, usually, one or more additional metals.

1) There are several important *physical properties of minerals.*

- a) *Crystal form* refers to the distinctive geometric arrangement of atoms or ions in the mineral. For a particular mineral, this depends on the attractive forces at the atomic level.
- b) *Cleavage* refers to the tendency of a mineral to split along preferred directions. In these directions, the attractive forces between ions or atoms are weak. If there are no weak directions, then the mineral just breaks in a random fashion.

c) *Specific gravity* is the density of the mineral divided by the density of water.
d) *Hardness* is a measure of how difficult it is to scratch the surface of the mineral. A standard 1-to-10 scale has been developed which consists of ten minerals, ranging from softest (talc, 1) to hardest (diamond, 10). Any given mineral can scratch all the ones below it on the list; it can be scratched by those above it.
e) *Streak* is the color a mineral leaves when it is rubbed against an unglazed white porcelain plate.

2) *Some noteworthy common minerals* are described below.

a) *Quartz*, SiO_2, has a hexagonal crystal form with no cleavage planes. It has a specific gravity of 2.65 and a hardness of 7. It is an important component of good optical glass.
b) *Feldspars* are silicates of either K or Ca (orthoclase) or Na, Ca, and Al (plagioclase). They have two cleavage planes at right angles to each other. Feldspars have a hardness of 6 and a specific gravity of 2.5-2.75. They are the most common minerals on earth and are used in the manufacture of porcelain.
c) *Mica* is a complex silicate with a very obvious cleavage plane; it breaks easily into sheets. It has a specific gravity of 2.9 but a hardness of only 2.5. Mica is used as an electrical insulator.
d) *Calcite*, $CaCO_3$, is a hexagonal crystal with three cleavage planes at angles of 75° to each other. It has a hardness of 3 and a density of 2.7. Lime for gardens and lawns is made from calcite.

Rocks

A *rock* is a mixture of minerals. It is not a compound because the minerals that comprise it retain their own properties. Rocks fall into three broad categories.

1) *Igneous rocks* are those that solidify from molten material called *magma*. Magma originates at great depths, where the temperature and pressure are very high. It is called *lava* if it reaches the surface in its molten state; rocks made from lava are called *extrusive*. If the lava solidifies below the surface, the rock is called *intrusive*.

The rate at which magma cools is very important. If it cools very rapidly, the rock has a glassy texture. If the cooling is slower, the mineral crystals have time to form grains with their characteristic shapes. The slower the cooling, the larger the grains. Sometimes the magma

INTRODUCTION TO GEOLOGY

solidifies while gases are escaping. The resultant rock then has numerous holes called *vesicles*.

2) *Sedimentary rocks* are those which are formed in layers, called *strata*. *Clastic sedimentary* rocks are formed when mineral grains are deposited and pressed together. *Nonclastic sedimentary* rocks do not start out with grains. Some precipitate out of solution. Others result when hard shells from dead microscopic marine animals collect on the sea floor. *Coal* forms when large quantities of dead plants accumulate in an oxygen-poor environment and do not decompose. Instead, they pile up and get compressed into rock.

3) *Metamorphic rocks* are igneous and sedimentary rocks that have been changed by intense heat and pressure. Although these conditions are not extreme enough to cause rocks to melt, they can cause small mineral grains to recrystallize into larger ones. The great pressure can also cause randomly oriented grains to line up perpendicular to the direction of the compressing force. In general, metamorphic rocks end up denser after these changes.

Soils

Soil consists of fine rock particles, moisture, and organic matter (humus), which together are able to support plant life. A cross section of a typical soil reveals three layers. The *surface soil* contains the most organic material. The layer just beneath, called the *subsoil*, is rich in soluble compounds that have been washed down from the surface soil by rain. Some organic matter is also present. Underneath these layers is the *parent material* which contains almost no organic matter at all. Different regions of the world have soils whose characteristics reflect the prevailing weather conditions.

1) *Pedocals* are characterized by the accumulation of $CaCO_3$ in the subsoil layer. This type of soil occurs in the dry regions of the western United States. Often a white powder, called *caliche*, is found on top of these soils. It results when water evaporates before it can wash away all the calcium carbonate. Hence, some crystals, which are white, remain.

2) In *pedalfers*, Al and Fe compounds accumulate in the subsoil. They are found in the eastern half of the United States.

3) *Laterites* are deep soils of the tropics that have high concentrations of aluminum and iron oxides. This is because other elements are washed

out by water. In addition, laterites have very little humus and are hard enough when dry to fabricate into bricks. They are poor soils for crop cultivation.

STREAMS

A *stream* is any body of water which flows in a channel. The term *river* is usually used to denote a large stream into which several smaller ones flow.

Stream Characteristics

1) The *base level* is the lowest level of the stream channel.
2) The *channel shape* is the shape of the cross section of the stream. Shape affects the amount of friction the walls of the channel can exert. The smaller the perimeter of the cross section for a given area, the less the friction. The most efficient shape in this regard is the semicircle.
3) *Discharge* is the volume of water flowing past a point each second.
4) The *gradient* is the decrease in altitude of the channel per unit length.
5) *Load* refers to the solid material that is swept along by the stream. The largest particle that can be carried is called the *competence*, a quantity that is proportional to the *square* of the velocity. *Capacity* is the maximum load that a stream can carry.

Stream Dynamics

In addition to erosion, to be discussed later, streams can transport and deposit solid material.

1) Most of a stream's load is *transported* either in suspension or in solution. Sometimes, material on the channel floor can roll or slip along. Occasionally, it gets lifted up, is carried along, and is dropped a few seconds later. This process is called *saltation*.
2) Material *deposited* by a stream is called *alluvium*. Deposition occurs when the stream slows down and loses competence. A *delta* is a triangular-shaped deposit and occurs in the region where a stream meets

INTRODUCTION TO GEOLOGY

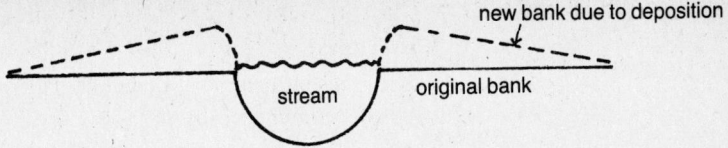

Fig. 21-1 Cross section of a natural levee

the ocean and slows down almost completely. An *alluvial fan* looks like a delta but is formed when a mountain stream reaches a flat plain. The sudden reduction in gradient causes a reduction in velocity and, hence, in competence. A *natural levee* is a wall of sediment along the banks of a stream, shown in figure 21-1. When a stream overflows, deposition occurs because the velocity of the overflow is relatively small. The points closest to the banks get the most sediment, giving the shape indicated in the diagram.

The Stream Valley

A *stream valley* is the region surrounding the stream on either side. It can be V-shaped, flat, or somewhere in between. A *drainage basin* is the land that channels water from rains into the stream. A *divide* is an imaginary line separating basins. The main features of a stream valley are summarized below. The evolution of a stream valley will be discussed later.

1) A *floodplain* is a flat, stream-valley floor.
2) A *meander* is a wide, sweeping curve in a stream that flows along a floodplain.
3) An *oxbow lake* is a meander that is separated from the stream.
4) *Rapids* are regions of a stream where the gradient changes abruptly. If the gradient has a sudden vertical drop, it is called a *waterfall*. Both of these features are characteristic of narrow valleys.

ICE AND WATER

In this section, we look at the major repositories of ice (glaciers) and water (oceans) on the earth's surface.

Glaciers

A *glacier* is a tremendous mass of ice that can move downhill. The ice is formed when a layer of snow gets compressed by the snow that has accumulated above it. About 10 percent of the earth's surface is covered by glaciers, either in the form of *valley glaciers* (confined to mountains) or *continental glaciers* (Antarctica or Greenland).

1) The *movement* of glaciers is slow and nonuniform. The center moves faster than the edge, as illustrated by figure 21-2. The line of o's represents stakes driven into the ice at a particular time. The same stakes are shown at a later time by the arc of x's. The lower layers of ice are the ones that actually flow; the top just gets carried along. The glacier gets larger if snow accumulates faster than it melts. If the reverse is true, the glacier gets smaller. At all times, however, the ice in the glacier flows forward.

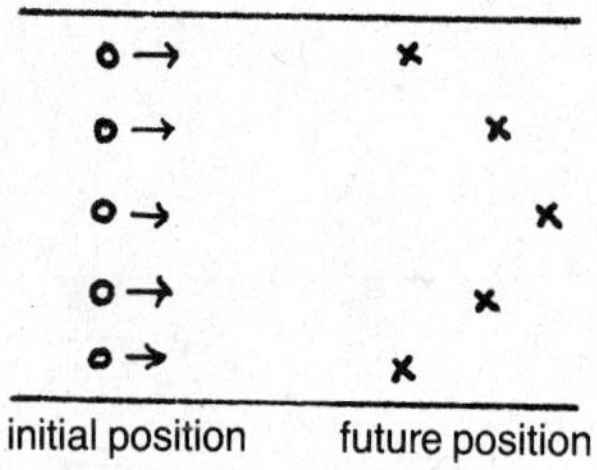

```
         o →             x
         o →               x
         o →                 x
         o →               x
         o →             x
      initial position  future position
```

Fig. 21-2   Glacial motion (top view)

2) *Moraines* are piles of debris deposited by the glacier. There are numerous kinds. *Lateral moraine* is a ridge left on either side of the glacier after it melts. *End moraine* is deposited at the front end (terminus) of the glacier. It forms when the glacier neither advances nor recedes. Since the ice has kept moving forward, debris accumulates at the terminus. If the glacier advances to a new position and becomes stagnant there, another end moraine results. The *terminal moraine* is the end moraine associated with the most advanced position of the terminus. If the glacier recedes, it leaves a layer of *ground moraine* in its wake. If it becomes stagnant after receding, the resulting end moraine is called a *recessional moraine*.

3) The *meltwater* at the terminus of a glacier carries sediment which eventually gets deposited in front of the end moraine. The area of this deposition is called an *outwash plain*. If a chunk of ice gets stuck in these deposits, a depression called a *kettle* forms; this fills with water when the ice melts.

INTRODUCTION TO GEOLOGY

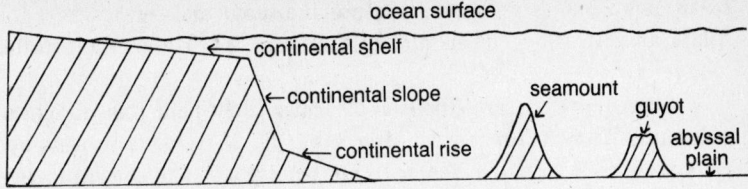

Fig. 21-3 The sea floor.

Oceans

An *ocean* is a huge body of salt water. Technically, there is only one ocean, which covers about two-thirds of the earth's surface. However, it is customary to call different regions by familiar names such as Atlantic and Pacific. The term *sea* is usually reserved for bodies of salt water smaller than oceans.

1) A *continental margin* consists of three regions. These and other features of the ocean crust are shown in figure 21-3. The *continental shelf* slopes very gently from the end of the land. It is really an extension of land that happens to be under water. However, it is a region rich in resources and fishing. At some point, which varies with locality, the gradient increases sharply, marking the beginning of the *continental slope*. This region, which slopes about 70 m/km, eases off to less than 10 m/km to form the *continental rise*.

2) An *ocean basin* begins where the margin leaves off. *Abyssal plains* are areas which are extremely level. *Seamounts* are isolated undersea volcanoes with heights of at least 1 km. *Guyots* are flat-topped seamounts. *Oceanic ridges* are mountain chains that stretch across the ocean floor.

3) *Seawater* is water with numerous salts dissolved in it. The most abundant salt is sodium chloride, averaging about 23 parts per thousand (23 o/oo). Magnesium chloride averages about 5 o/oo while sodium sulfate is about 4 o/oo. Everything else is present in amounts less than 1 o/oo. The *salinity* is the proportion of the total salt to pure water. Although the average is about 35 o/oo, it can vary from 2 o/oo (Baltic Sea) to 42 o/oo (Persian Gulf), depending on evaporation, precipitation, and fresh water flow from rivers.

Solute is provided to the ocean by rivers, which empty into the ocean, and by volcanoes, which emit soluble gases. However, plants and animals use some of these substances in their life processes. Also, solute can be precipitated out as sediment. Because the input and output rates for solute are the same, the salinity of the ocean is fairly constant in the long term.

4) *Ocean currents* are important because they help transport heat poleward. *Surface currents* are driven by the wind. Since moving objects veer to the right in the northern hemisphere, the current in any given layer (down to about 100 m) is always to the right of the current in the layer above. Naturally, the deeper layers have smaller velocities due to frictional losses. Figure 21-4 shows this symbolically. Each vector represents the current of a particular layer. Larger numbers are for deeper layers. Note that the lowest layer in the figure is the result of so many successive deflections to the right that it is actually in a direction opposite the surface current. This phenomenon is called the *Ekman spiral* (a line connecting the tips of the vectors forms a spiral). *Deep currents* are usually the result of density differences: denser water sinks with respect to lighter water. Density increases as temperature decreases; it also increases as salinity increases. Hence, changes in temperature and salinity can start currents moving. *Upwelling* (figure 21-5) is the rising to the surface of cold, deep water. Along the coast of California, for example, surface winds are such that the upper layers of water move away from the coast. These waters are replaced by slow upwelling from the bottom. Not only does this cool the local environment, it brings up concentrations of nutrients which support large fish populations.

EROSION, WEATHERING, AND MASS WASTING

Erosion

Erosion is a process whereby rocks are broken down and the resulting debris gets carried away.

1) The *evolution of a stream valley* (figures 21-6 through 21-8) is a direct result of *stream erosion*.

   a) In the *youthful stage,* the stream erodes its channel downward until it reaches base level. The channel is V-shaped during this process.

INTRODUCTION TO GEOLOGY 255

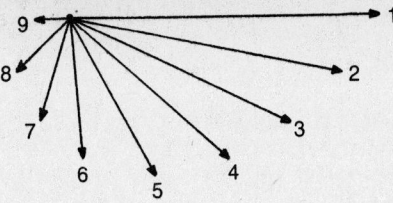

Fig. 21-4 Ekman spiral

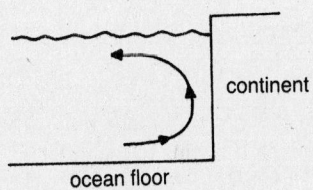

Fig. 21-5 Upwelling

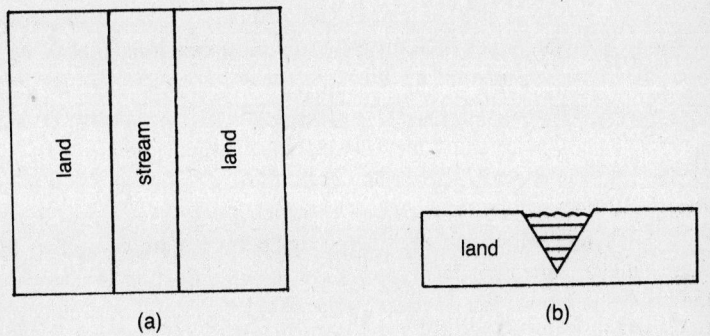

Fig. 21-6 (*a*) Top view of youthful stream valley. (*b*) Cross section of youthful stream valley.

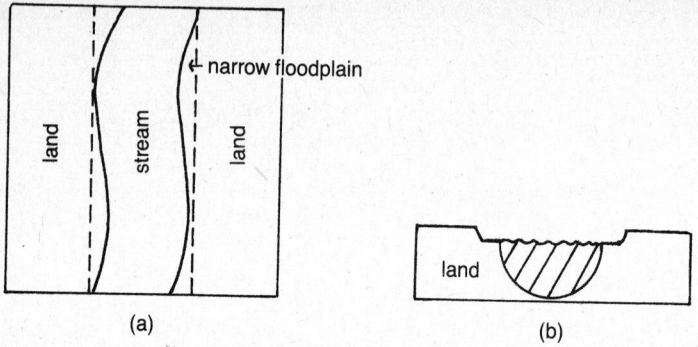

Fig. 21-7 (a) Top view of mature stream valley. (b) Cross section of mature stream valley.

b) In the *mature stage,* the stream cuts sideways, carving out a floodplain on which it begins to meander.

c) In the valley's *old age,* the meandering is extreme, with the floodplain several times wider than the lateral extent of the meanders. Oxbow lakes are often present in this stage.

2) *Coastlines* are shaped by *wave erosion. Wave-cut cliffs* result when waves erode the base of the land, producing the notched shape of figure 21-9. When the overhang crumbles, a *wave-cut platform* results. A piece of land protruding into the sea is eroded on both sides by the waves. The softer areas disintegrate first, producing caves. If the erosion is severe enough, two caves on opposite sides can join, producing an *arch.* Eventually, the roof of the arch caves in, leaving upright structures called *stacks.*

3) *Glaciers* cause erosion by either carrying away loose pieces of rock or grinding the surface with bits of rock that they normally drag along. Relatively flat regions show *striations,* which are grooves scored in rock by the glaciers. Mountain regions occupied by ice are gouged so that their valleys, called *glacial troughs,* are deep, wide, and U-shaped. At the head of a trough is a *cirque,* a basin where snow and ice accumulate. If a particular mountain is carved out on all sides by cirques, the result

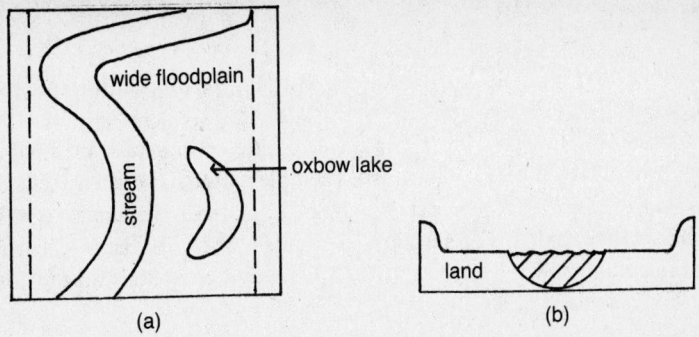

Fig. 21-8 (a) Top view of stream valley in old age. (b) Cross section of stream valley in old age.

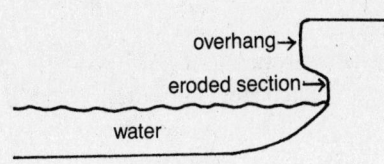

Fig. 21-9 Wave-cut cliff

is a sharp peak called a *horn*. The famed Matterhorn in the Swiss Alps is an example. If the valley is near the coast and gets flooded by the sea when the glacier recedes, the resulting form is called a *fiord*.

4) *Wind* is an agent of erosion in dry areas. In these areas, there is very little moisture or vegetation to hold the soil together. Wind can cause abrasion by blowing sand and pebbles against rock formations. It can also lift loose material and blow it away, a process called *deflation*. The material it lifts is fine-grained, leaving behind a coarse-grained surface called *desert pavement*.

Weathering

Weathering refers to the disintegration or decomposition of rock material.

1) *Chemical weathering* alters the structure of minerals. In this process, water is the vehicle that carries soluble material, which attacks the rock. *Dissolved oxygen* can combine with other elements to form oxides. Iron oxide, for instance, is formed when the iron in certain rocks combines with oxygen. *Carbon dioxide* reacts with water to form carbonic acid:

$$CO_2 + H_2O \rightarrow H_2CO_3$$

This acid attacks minerals such as feldspar.

2) *Mechanical weathering* causes rock to break into smaller fragments. In *frost wedging*, water seeps into cracks and expands upon freezing. The force of expansion breaks the rock apart. *Exfoliation* is the breaking loose of layers of rock which conform approximately to the shape of the rock. *Jointing* is the cracking of rock due to contraction, such as when magma cools.

Mechanical and chemical weathering are linked by an important fact: mechanical weathering increases the surface area so that chemical weathering can take place at an increased rate.

Mass Wasting

Mass wasting is the downward movement of rock and soil due to the force of gravity.

1) *Slow* movements are the most common. *Creep* is the slow down-

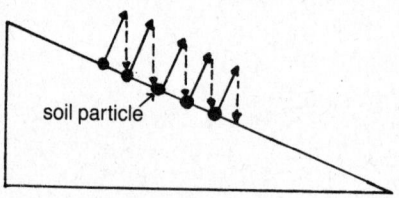

Fig. 21–10 Creep. Successive positions of a particle of soil

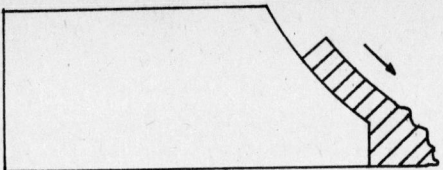

Fig. 21-11 Slump

ward movement of soil. It results when moist soil freezes and thaws. Figure 21-10 shows how the soil moves. During freezing, expansion causes the soil to move perpendicular to the surface. When it thaws, gravity pulls it straight downward. *Solifluction* is creep that occurs in frigid zones. Only a thin layer of soil thaws during the warm weather. The water in it cannot seep farther into the ground because the ground is permanently frozen. Hence, it remains in the thawed layer, which flows downhill.

2) *Fast* movements are the most dramatic. Rockslides occur when chunks of loose rock fall down the side of a steep grade. *Slump* occurs when a mass of earth slides down along a curved surface, as shown in figure 21-11. In *earthflow,* a quantity of earth simply slides down a hill for a limited distance. If the earth is particularly wet, it flows in a more fluidlike fashion and is called *mudflow*.

22
The Earth's Interior

In this chapter, we will investigate various features and characteristics of the region beneath the earth's surface.

EARTHQUAKES

An *earthquake* is a trembling of the ground due to a sudden release of mechanical energy.

Rock Deformation

1) *Stress* refers to forces that act on a rock. Sharp, sudden stresses can break a rock into pieces. Weaker stresses applied over a long period of time, however, can cause a rock to bend.

2) *Elastic strain* is the bending of rocks under the influence of stress. If the strain is within a certain limit, a rock can revert to its original shape without breaking when the stress is removed.

3) *Elastic rebound* is the snapping action of a rock when the limit of elastic strain is exceeded.

Figure 22-1 shows how a rock formation bends and breaks when a stress is applied and the elastic limit is exceeded. Note that the broken pieces assume their original shapes after snapping. It is the snapping action that provides the energy for earthquakes.

Faults

A fault is a crack in the earth's surface along which sections of the crust get displaced.

THE EARTH'S INTERIOR

1) A *strike slip fault* (figure 22-2) is a fracture along which horizontal movements take place.

2) A *normal fault* (figure 22-3a) is a nonvertical fracture where the rock above the fracture (A) moves *downward*.

3) A *reverse fault* (figure 22-3b) is a nonvertical fracture where the rock above the fault moves *upward*.

If the motion along a fault is gradual and steady, no earthquake will take place. It is only when the sections on both sides get "stuck," then suddenly snap, that an earthquake occurs.

Location, Magnitude, and Prediction

1) The *location* of an earthquake is called the *focus*. This is always below the surface. The *epicenter* is a point on the surface directly above the focus.

2) The *magnitude* of an earthquake is a measure of how strong it is. The *Richter* scale is a measure of the actual energy released. The weakest quake has a value of 0. Each increase of 1 on this scale represents a tenfold increase of energy. A rating of 5 is about equal to the first atomic bomb, while 7 is a major quake. The largest rating on record is 8.6. The modified *Mercalli* scale attempts to assess quakes in terms of everyday phenomena. For example, a particular quake may be described as rumbling floors or moving heavy furniture about. This scale has twelve categories of such descriptions. It does not represent actual energy released, however. This is because a nearby quake that is weak could have greater effects than a strong one that is distant.

3) *Prediction* of earthquakes centers on studying phenomena that precede the quake itself. Scientists have discovered that slow movement along some faults comes to a halt before a major quake. They have also found that very small quakes, called *microquakes*, precede the main one. By monitoring these events, scientists hope to uncover patterns that will ultimately lead to successful prediction.

Seismology

Seismology is the study of waves that are sent out by earthquakes.

1) *Surface waves* travel across the surface like water waves. They are the ones that do the damage characteristic of earthquakes. *Primary (P) waves* are longitudinal and travel through the earth. They can penetrate

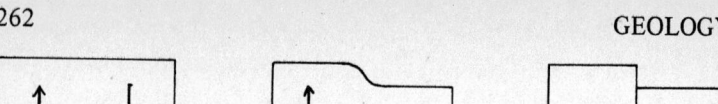

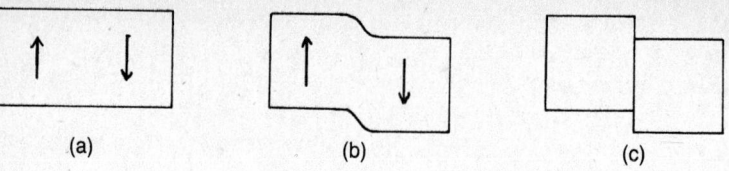

Fig. 22-1 Deformation and snapping of rock. Arrows indicate directions of stresses.

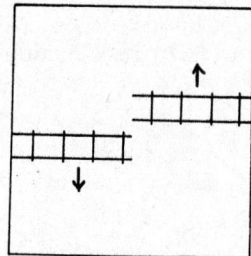

Fig. 22-2 A strike slip fault. Top view of tracks indicates relative movement.

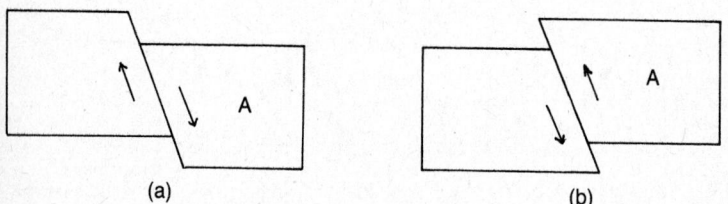

Fig. 22-3 (*a*) Normal fault. (*b*) Reverse fault.

THE EARTH'S INTERIOR

solids, liquids, and gases. *Secondary (S) waves* are transverse and also travel through the earth. However, they can penetrate only solids. Also, they travel more slowly than P waves.

2) The *distance to the epicenter* is found by measuring the time difference between the arrival of the P and S waves at a monitoring station. This can be done because the longer the waves are in transit, the farther behind the slower S wave drops, increasing the delay in its arrival at the station. The *location of the epicenter* is found by using three monitoring stations as shown in figure 22-4. Each station measures the distance to the epicenter and draws a circle, centered at the station, with a radius equal to the measured distance. The intersection of the circles is the epicenter.

3) The *structure of the earth's interior* is revealed by the behavior of P and S waves. Figure 22-5 shows the paths of waves as they go through the earth. Abrupt changes in the path are due to refraction at the boundary between two layers. There is a *shadow zone* where no earthquake waves can be detected. This is between $104°$ and $143°$ from the epicenter on both sides. Yet beyond $143°$, P waves can be detected. This is because the core has two layers. The *inner core* is solid. The *outer core*, however, is liquid. The liquid outer core prevents S waves from getting beyond $104°$. The refraction of P waves at the mantle-core interface is the reason why even P waves do not show up in the shadow zone.

4) The *seismograph* detects earthquakes. Figure 22-6 shows the principle behind it. A slowly rotating drum is attached to the base of the seismograph. This, in turn, is firmly fixed to the earth. The pen is attached to a heavy weight hung from the seismograph by a wire. A tremor shakes the entire seismograph *except* the heavy weight; although the wire gets wiggled around, the huge mass of the hanging weight is hardly budged. The rotating drum, which also shakes, records the quake.

IGNEOUS ACTIVITY

Igneous activity refers to any process involving molten rock. Because of radioactive decay in the upper mantle, heat is produced which can partially melt rocks. This is believed to be responsible for the production of magma.

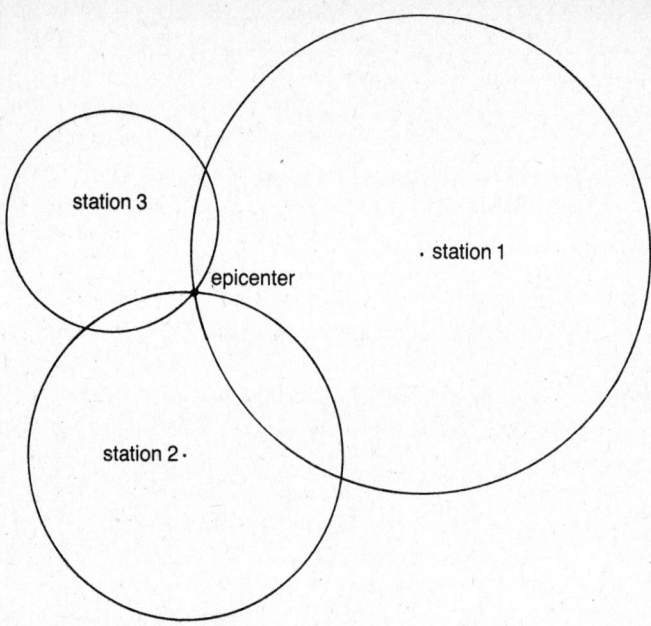

Fig. 22–4 Location of earthquake epicenter. Each station draws a circle whose radius equals its distance from the earthquake. The intersection of the three circles is the epicenter.

Volcanoes

Volcanic activity is the eruption of magma through a crack in the earth's surface. Most volcanic activity is in the form of relatively quiet eruptions through narrow cracks called *fissures*. The most spectacular eruptions occur through relatively straight passageways called *vents*. A *volcano* is an accumulation of material which is the result of numerous eruptions through a single vent.

1) *Ejected material* is quite varied. Lava flows can be ropy or jagged in appearance. These are given the Hawaiian names *pahoehoe* and *aa*. *Pyroclastics* are fragments of various sizes that are ejected explosively from the vent. These vary in size from *dust* to large, streamlined *volcanic bombs*. A *nuee ardente* is a thick, potentially devastating cloud of hot gas and incandescent particles.

2) There are three *types of volcanoes*. The *shield volcano* is a broad

THE EARTH'S INTERIOR

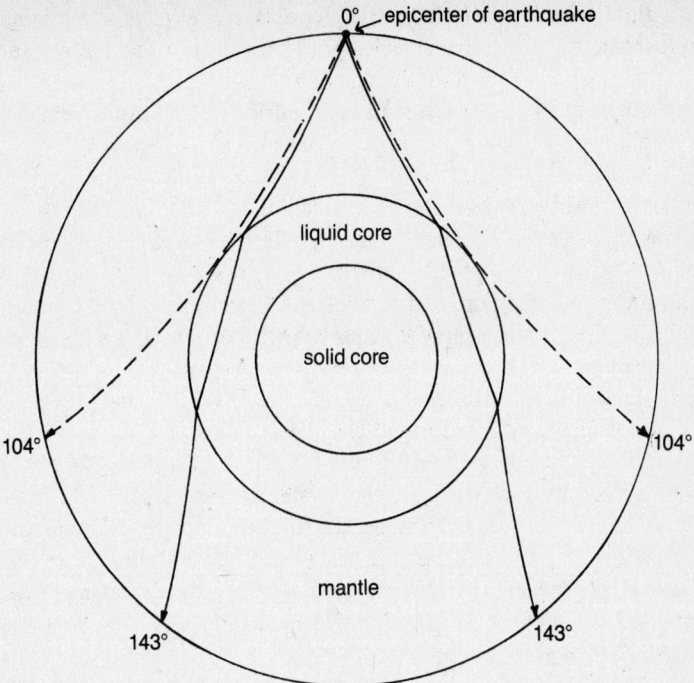

Fig. 22-5 Earthquake wave paths. S-waves (dashed line) cannot get beyond 104° due to their inability to penetrate liquids. P-waves (solid line) penetrate liquids but, due to refraction, are absent from 104° to 143°.

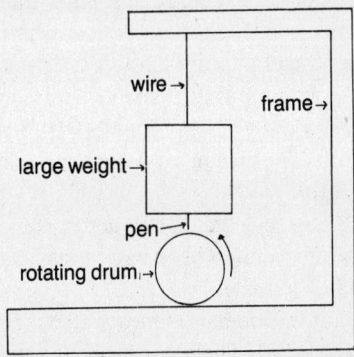

Fig. 22-6 Seismograph. The frame, drum, and wire all shake, but the weight (hence, the pen) does not.

lava flow with a relatively gentle slope. A *cinder cone* is a mound of pyroclastics with a rather steep slope. These cones, of all the volcanoes, are the most susceptible to erosion. Sometimes, erosion is so extensive that the only thing left is a mass of solidified magma in the vent, called a *volcanic neck*. A *composite cone* has alternating layers of lava flow and pyroclastics.

Most volcanoes have a *crater*, a widening of the neck at the summit. Some have *calderas*, which are unusually large craters. These form when a filled crater plugs up the vent and prevents magma from escaping. The pressure builds up, and the volcano literally blows its top. Sometimes a filled crater simply collapses downward, leaving a much larger opening in its place.

3) *Volcanic gas* is dissolved in the magma and makes up a few percent of the total volcanic eruption. The gas is important for two reasons. First, the atmosphere was produced from volcanic gas; so were the oceans (volcanic gas is 70 percent water vapor). Second, the way in which the gas escapes determines the kind of eruption. If the magma is relatively thin, the gas escapes easily and the eruption is relatively quiet. If the magma is thick, the gas has a hard time escaping. Pressure can build up, making the eruption violent.

Intrusive Formations

As we saw in the last chapter, magma can solidify underground and produce intrusive igneous rocks. Here we look at some of the large-scale intrusive formations, aided by figure 22-7.

1) A *batholith* is an enormous intrusive structure located deep within the earth.

2) A *sill* is a sheetlike intrusion of magma which is *parallel* to the layers into which it intrudes.

3) A *dike* is like a sill except that it cuts *across* the layers.

4) A *laccolith* is an intrusion whose base is parallel to the layers but whose top is shaped like a dome.

5) A *vein* is a deposit of minerals from a watery solution that remains after magma has crystalized. The solution penetrates thin fractures of rocks. Veins are the structures that are mined for important minerals.

THE EARTH'S INTERIOR

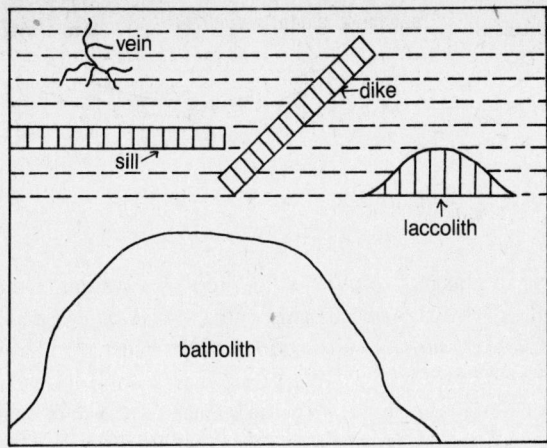

Fig. 22–7 Intrusive features

MOUNTAINS

A *mountain* is a large region of the crust that extends significantly above its surroundings. Mountains are the result of enormous forces that have uplifted portions of the crust. The evidence for uplift comes, for instance, from marine fossil remains found atop tall mountains. This means that the part of the crust containing the fossil was once underwater. A check of various shoreline features around the world shows that this is not the result of a drop in sea level. Hence, uplift has taken place.

Mountain Types

1) *Folded mountains* result from the uplifting and bending of layers of rock. Figure 22-8 shows a cross section of rock layers that have been folded three ways. The folds look like waves. The crestlike shapes are called *anticlines,* while the troughlike shapes are called *synclines*. A *symmetrical fold* has the same shape on either side of the peak; an *asymmetrical fold* has slightly different shapes on each side. In an *overturned fold*, the peak "leans" over to one side.

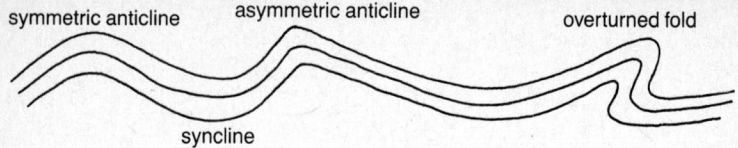

Fig. 22–8 Folds

Folded mountains make up the major mountain systems in the world. These include the Alps, Appalachians, Himalayas, Rockies, and Urals.

2) *Fault block mountains* are sections of the crust that have normal faults as boundaries. Figure 22-9 shows a cross section of terrain where faulting has occurred, along with the names of the two shapes that result. The depression is called a *graben,* while the raised portion is called a *horst.* Large-scale examples of horsts are the Sierra Nevadas and Tetons.

3) *Volcanic mountains* are volcanoes, which we have discussed. Examples are Mounts Fuji, Hood, Ranier, and Shasta. The most recently active volcano has been Mount Saint Helens, in Washington.

4) *Domed mountains* are regions where the crust has been pushed upward. A bulge in the surface can occur, for instance, because of an intrusion such as laccolith. The Black Hills of South Dakota are domed mountains.

Isostasy

Isostasy is the concept that the crust floats in the mantle. The top of the mantle is a region of partial melting, giving it the ability to flow

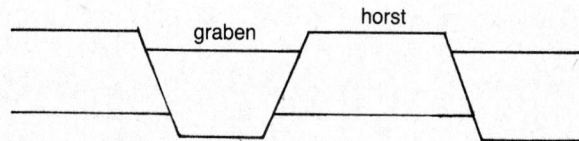

Fig. 22–9 Horst and graben

THE EARTH'S INTERIOR

somewhat. The floating crust is analogous to the way that blocks of wood float in water. Let us pursue this analogy.

We know that the amount of water displaced depends on the mass of the floating object. Assume that a block of wood has a density such that it floats half submerged. Consider another block twice as tall but otherwise identical to the first. Since it has the same density, it also floats half submerged. However, it is taller, so it not only sticks out of the water more, it protrudes deeper into it (figure 22-10). Now, what happens if we cut off a slice from the top of the block? The buoyant force causes a readjustment of the floating position; it bobs up so that it continues to float half submerged. This means that less protrudes into the water than before the cut. Similarly, if we place a weight on the block, it floats with more material sticking into the water.

If isostasy is correct, then mountains must extend both above and below the surface. This is in fact the case, as various investigations have shown. Furthermore, an enormous weight placed on the crust should cause some sinking. This is exactly what happened to the area around Lake Mead when it was flooded to get Hoover Dam started. Just the opposite has occurred in parts of Scandinavia, where heavy glaciers once stood. Their retreat eased a heavy burden from the crust, which is still adjusting by rising slowly upward.

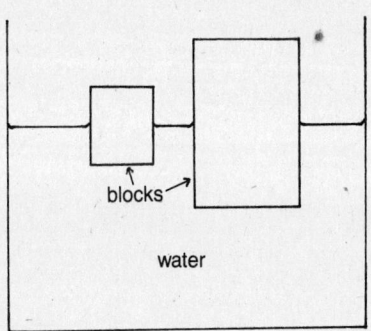

Fig. 22–10 Isostasy model. The taller block protrudes deeper into the water.

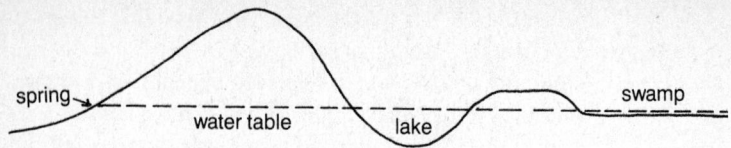

Fig. 22-11 Features associated with groundwater

GROUNDWATER

Groundwater is the water that saturates the ground. Usually this occurs under the surface in a porous zone that follows approximately the contours of the surface. The *water table* is the top of the saturated zone. *Permeability* is a measure of how easy it is for water to flow in the saturated rock. An *aquifer* is a highly permeable rock layer. Even in an aquifer, however, water flows very slowly (a few meters a day).

Groundwater and Topography

A number of geological features are associated with groundwater. (See figures 22-11 and 22-12.)

1) A *swamp* is a place where the water table is just at the surface.

2) A *lake* results if there is a depression in the surface which dips below the water table.

3) A *spring* is a place where the water table intersects the surface.

4) A *cavern* is an underground hole that results when a section of a limestone deposit is dissolved by groundwater.

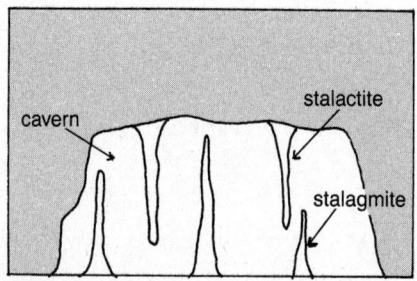

Fig. 22-12 Cavern features

THE EARTH'S INTERIOR

5) A *stalactite* forms when water drips from the roof of a cavern and evaporates, leaving behind a deposit of calcite hanging from the roof.

6) A *stalagmite* forms when water drops that hit the floor evaporate, also leaving behind calcite deposits, which build upward.

Wells

1) A *well* is a hole that extends from the surface to below the water table. Groundwater collects at the bottom and must be pumped to the surface in order to be used. If the rate of pumping exceeds the rate at which groundwater is replenished, the water table drops. This can affect neighboring wells, which can go dry if they are not dug deep enough.

2) In an *artesian well*, the water level rises naturally to a level above the surrounding water table. Figure 22-13 shows how this is possible. An aquifer is situated between two impermeable rock layers. One end (A) is at the surface and can absorb rainwater up to the water table (dashed line). If a hole is drilled at B, water fills the well almost up to the level of the dashed line. This is because atmospheric pressure at A pushes water up the well until the weight of the water balances the upward force. Due to friction, however, the water does not rise as high as is ideally possible. Water in this well must also be pumped up, in order to be useful. At well C, however, the level to which the water rises is above the surface, so no pumping is required.

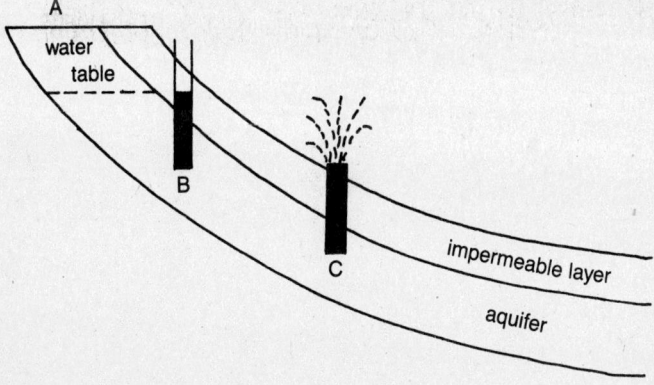

Fig. 22–13 Artesian wells

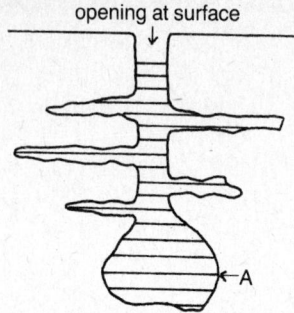

Fig. 22–14 Geyser. The water at A is under great pressure due to the water above it.

Geysers

A *geyser* is a natural opening in the ground from which heated groundwater intermittently erupts. There are a number of regions in the world where magma is relatively close to the surface. This is why the groundwater can be heated. Figure 22-14 shows a model of a geyser. Water in the lower part of the geyser is under great pressure because of all the weight on top of it. This means its boiling point is very high, say 200°C. Let us assume that the water at the bottom gets heated to 195°C. The water expands, forcing some water at the top out of the hole. This reduces the pressure at the bottom which, in turn, reduces the boiling point. When the new boiling point drops below the existing temperature, the water boils suddenly and spews forth from the hole. Eventually, groundwater refills the hole, gets heated, and undergoes the same process over again.

23
Evolution of the Earth

In this chapter, we will first discuss a relatively new theory called plate tectonics, which seems able to account for the earth's physical evolution over the past 200 million years. Then, after learning how scientists date ancient rocks, we will briefly review the major time intervals of earth history.

PLATE TECTONICS

Plate tectonics is a theory that sections of the earth's crust, called *plates*, move relative to one another as they float in the upper mantle (also called the *aesthenosphere*). All land was once combined in a single supercontinent called *Pangaea*. The northern part (Europe, Asia, and North America) is called *Laurasia*. The southern part (Africa, Antartica, Australia, India, and South America) is called *Gondwanaland*. The plates separated and the various continents spread apart to their present positions. According to plate tectonics, the motion of a given plate boundary with respect to another is responsible for the major geologic activity that we have studied.

Plate Boundaries

1) *Convergent boundaries* (figure 23-1a) are regions where two plates come together. *Subduction* is the sinking of an ocean plate into the aesthenosphere when it converges on another plate.

2) *Divergent boundaries* (figure 23-1b) are regions where two plates move away from each other.

3) *Transform fault boundaries* (figure 23-2) are regions where plates slip by each other sideways.

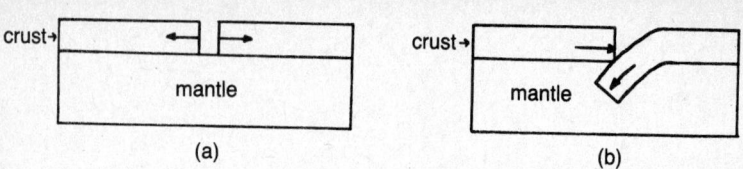

Fig. 23–1 (a) Convergent boundaries with subduction. (b) Divergent boundaries.

Evidence for the Plate Model

1) *Sea-floor spreading* is the most significant evidence for the plate model. Measurements have been made of the age of the ocean floor in the vicinity of the Mid-Atlantic ridge. These measurements show that the age of the crust increases as one moves away from the ridge on either side. This shows that new oceanic crust must form at the ridge and spread out.

2) *Paleomagnetism*, explained here, shows that the spreading sea floor was once molten. When a rock is above a particular high temperature, called the *Curie point*, the magnetic moments of certain component minerals line up with the earth's field. Therefore, if the earth's field direction changes, so does the direction of the magnetic moments. When the rock cools to below the Curie point, these magnetic moments remain fixed in direction, no longer affected by the earth's magnetic field. Now, it is known that the earth's field has reversed itself a number of times. These reversals are recorded in the ocean floor, as shown in figure 23-3. Note that as one proceeds outward from the ridge on both sides, the magnetic moments reverse direction. The pattern is symmetric about the ridge. The fact that this pattern appears at all is significant. It means that the ocean floor did not cool all at once. If it

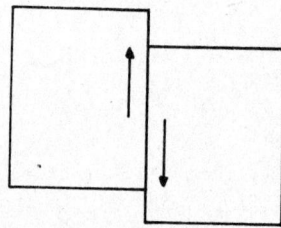

Fig. 23–2 Transform fault boundaries (top view)

EVOLUTION OF THE EARTH

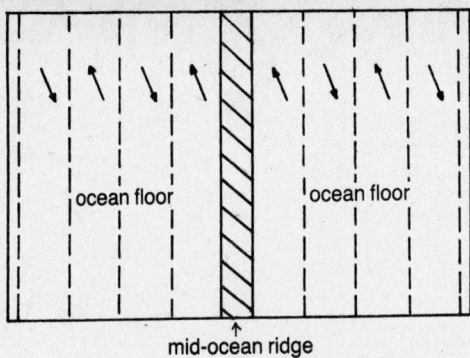

Fig. 23-3 Paleomagnetic reversals. Arrows show direction of earth's magnetic field at the time each section of the ocean floor cooled.

did, then all the moments would point in the same direction. Different sections must have cooled (to below the Curie point) at different times, with the earth's field direction at the time permanently frozen in. The only place where hot material cools is at the ridge, which is the youngest part of the ocean floor. Thus, we conclude that molten material from the mantle rises to the surface until it reaches the ridge, where it cools. It then becomes the newest part of the sea floor and spreads out, carrying with it a permanent record of the direction of the earth's magnetic field at the time of cooling.

3) *Fossils* also provide evidence for the plate theory. The fossil *Glissopteris* has been discovered in places like Europe, India, and Antarctica. This creature could not have lived in places with such disparate conditions. Hence, these areas must all have been in a similar location.

Applications of Plate Tectonics

1) *Earthquakes* occur around plate boundaries because they are regions where one section of the crust suddenly slips with respect to another. Earthquakes with deep foci (deeper than 300 km) occur only in subduction zones, while those with foci greater than 700 km rarely occur at all. Figure 23-4 shows how plate tectonics explains this. We expect earthquake foci to be found anyplace where the crust moves. A

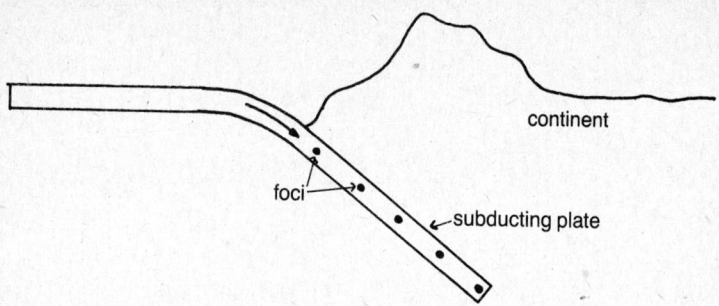

Fig. 23–4 Location of earthquake foci

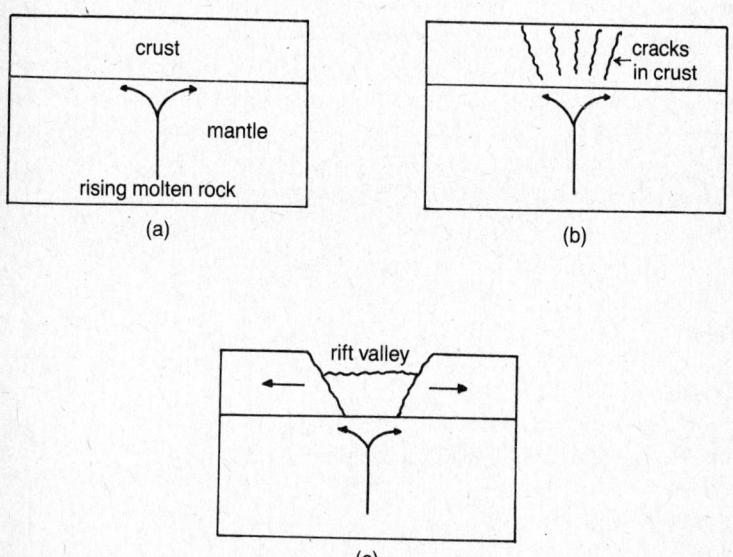

Fig. 23–5 Formation of a rift valley

EVOLUTION OF THE EARTH

subduction zone is the only place where moving crust is found at such great depths. Hence, the subduction zone is the only possible place for deep-focus earthquakes. Furthermore, when the crust reaches the aesthenosphere, it is no longer rigid and cannot produce the snap needed for an earthquake. That is why earthquake foci are rarely found below 700 km.

2) A *rift valley* has its beginnings when molten rock rises underneath continental crust. Figure 23-5 shows what happens. The rising molten rock initially spreads outward when it reaches the crust. Soon, cracks develop in the crust. Finally, the broken crust gets pulled apart by the currents of molten rock, leaving a valley in the divergent zone. It is believed that this is the mechanism by which a continent initially breaks up.

3) *Volcanic activity* that occurs in divergent and subduction zones fits well into the plate model. In a divergent zone, magma simply rises to fill a space. Sometimes it spreads slowly outward; sometimes it erupts violently. In a subduction zone, the sinking crust eventually melts, generating magma which gets forced upward through cracks. Volcanoes that occur away from plate boundaries are not easy to explain. They fit the plate model only if it is assumed that they are the beginning of a rift valley.

4) *Mountains* are best explained by the plate theory. In one case, sedimentary layers can build up on the continental shelf and continental rise. This accumulation is called a *geosyncline* (figure 23-6). If the edge of the continent becomes a subduction zone, the geosyncline gets deformed and uplifted, producing a folded mountain range at the edge of a continent. Folding does not have to occur this way, however. Two continental plates that converge do not subduct. Instead, they collide, with the resultant folding and uplifting producing spectacular mountains. The Himalayas, for example, resulted when the plate carrying India collided with Asia.

GEOLOGIC DATING

Absolute Dating

Absolute dating is the establishment of the precise age of an event or object. This process involves radioactive decay of a substance of known half-life, an idea described in Chapter 14 using carbon-14. Carbon-14 is

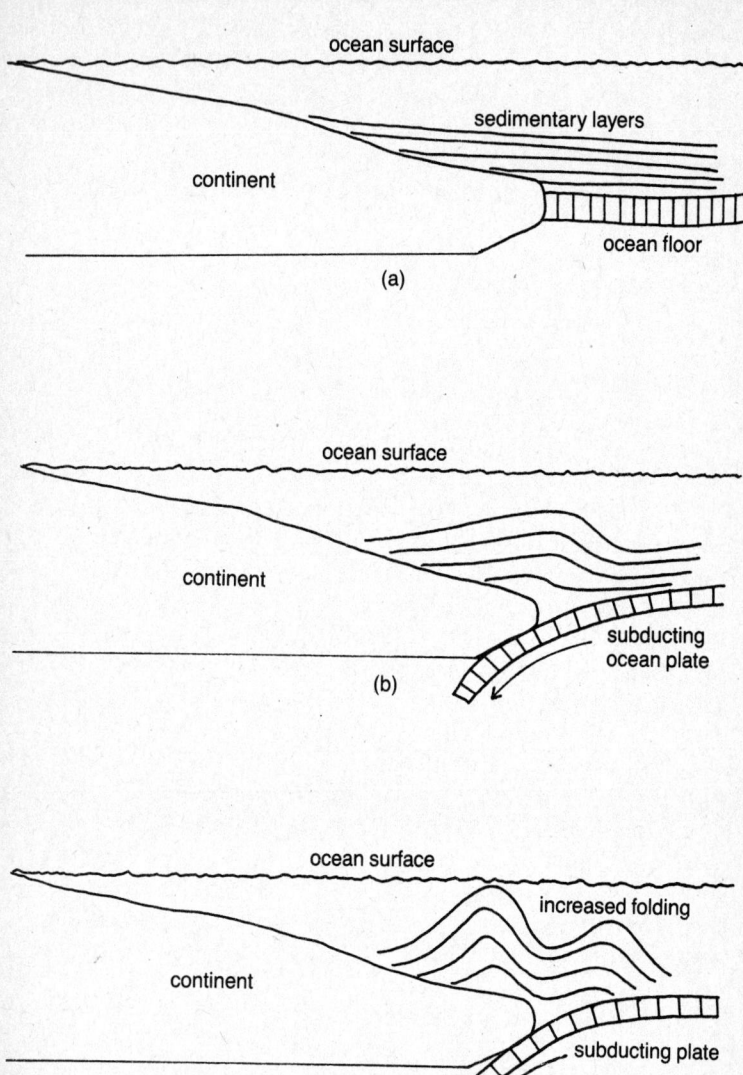

Fig. 23–6 Geosyncline. Sedimentary layers (*a*) begin to deform as subduction beings (*b*). Continued subduction results in upward folding of original layers (*c*).

EVOLUTION OF THE EARTH

good for dating objects that are at most 75,000 years old. This is because its half-life is only 5720 years; after 75,000 years, too small a fraction of carbon-14 remains for a reliable analysis to be made. For objects that are much older, other isotopes are used. Most useful is potassium-40, which decays to argon-40 with a half-life of 1.3×10^9 years. Using this isotope, one compares the amount of potassium to the amount of argon. Because of the long half-life, this method can be used to date very old rocks. Using sophisticated technology, very small amounts of argon-40 can be detected. Hence, the process can also be used to date rocks as young as 50,000 years.

Relative Dating

Relative dating is the placing of rocks in their proper age sequence. Three main ideas are used to do this.

1) The *principle of superposition* states that for a group of undisturbed sedimentary strata, each layer is younger than the one below it.

2) The *principle of crosscutting* states that features that cut through rocks are younger than the rocks themselves. For instance, the dike in figure 22-7 is younger than the layers it cuts.

3) *Correlation* is the process by which scientists determine the proper age relationship of widely scattered rocks. Consider a bed of 10 layers located at A, and another bed of 10 layers at B. We know that the top layer at A is younger than the second layer at A. Is the top layer at A younger than the second layer at B? We don't know until we are able to relate the ages of the two beds. To do this, scientists usually turn to fossils.

Fossils

A *fossil* is the preserved remains of an ancient organism. Preservation is made possible when organisms with hard parts such as shells or bones get buried soon after death.

Different creatures were particularly abundant during different time intervals. Hence, finding the same type of fossil in two disparate locations helps in the correlation process. Fossils can also indicate the type of environment that existed at the time. A particular plant or animal, for instance, may have been able to survive only mild temperatures.

Hence, it can be inferred that the rocks containing the fossil were laid down when the climate was mild.

GEOLOGIC TIME

The earth is known to be about 4.5 billion years old. This enormous span of time is divided into intervals called *eras*. Eras are divided into *periods*, which are further divided into *epochs*. Most of what we know about the earth's history is confined to the last 600 million years.

Precambrian Time *(> 600 million years ago)*

Although Precambrian rocks can be found in eastern Canada and at the base of the Grand Canyon, they contain very few fossils. Thus, despite the fact that this interval covers 86 percent of geologic time, the record of events here is very sparse.

We do know that simple life (invertebrates without shells) began in this era. The initial temperature of the earth was warmer than now. Subsequent cooling undoubtedly produced rains which carried down dissolved gases containing the elements necessary for life: carbon, oxygen, hydrogen, and nitrogen. Laboratory studies have shown that an electric discharge passed through a mixture of methane, ammonia, and water can produce complex molecules such as amino acids. It is probable that simple life on earth formed in a similar manner, with lightning providing the electrical input.

The Paleozoic Era *(225-600 million years ago)*

1) *Geologically,* the Paleozoic era began quietly, with much of the North American continent under water. Toward the end of the era, however, much crustal uplift took place, with the result that only the edges of the continent were under shallow seas. The organic material that ultimately resulted in the Appalachian coalfields was laid down at this time.

2) *Biologically,* this era marked the beginning of complex forms of life: fish, amphibians, reptiles, and extensive forests. These thrived on the uniformly mild climate that prevailed over the continents. Toward

EVOLUTION OF THE EARTH

the end of the era, the environment changed rapidly, with climatic conditions turning less favorable. Many species became extinct; only those whose evolutionary changes were also rapid could survive.

The Mesozoic Era *(65-225 million years ago)*

1) *Geologically,* the Mesozoic era was noteworthy for the breakup of Laurasia and Gondwanaland. Initially, the North American continent was above water. Sometime during the Jurassic period, however, a shallow sea pushed into the area of the western, plains, and Gulf coast states. By the end of the era, crustal uplift had occurred and the continent was once again above water.

2) *Biologically,* the era was characterized by an abundance of reptiles, especially dinosaurs. In addition, birds and small mammals appeared. Since they were warm-blooded and cared for their young, they were particularly well suited to survive the drastic climate changes that occurred at the end of the era.

The Cenozoic Era *(present)*

1) *Geologically,* the Cenozoic era has been characterized by crustal disturbances such as uplift and volcanoes. The *Ice Age* occurred during this era, with the last great ice sheet retreating about 30,000 years ago. This was a series of four advances and retreats of glacial ice. Every part of the northern border of the United States was invaded by glaciers; some glaciers got as far south as Kentucky.

2) *Biologically,* the Cenozoic era has been highlighted by the rise of a great variety of mammals such as horses, apes, camels, and, most notably, man. Remains of human creatures date from the Ice Age. According to fossil evidence, however, the evolutionary step up from the ape was probably made in the Pliocene epoch, which lasted from about 7 million years ago to about 2 million years ago. The radical changes of climate during the Ice Age were particularly hard on living things. Today, there is a smaller variety of mammals than before the Ice Age.

24
Pollution

In this chapter, we will summarize the major problems involved in the contamination of our air and water.

AIR POLLUTION

Pollutants

The Environmental Protection Agency estimates that about 2.5×10^8 tons of pollutants are added to the atmosphere every year.

1) *Carbon monoxide* is produced when the hydrocarbons that comprise fossil fuels are not completely burned. Normally, the combustion of a hydrocarbon results in H_2O and CO_2. During incomplete combustion, not enough oxygen is available to put two on each carbon, so CO forms instead of CO_2. CO can be a fatal blood poison. *Unburned hydrocarbons* are another result of incomplete combustion of fossil fuels.

2) *Oxides of nitrogen* (often abbreviated NO_x) are produced when the relatively inactive nitrogen in the air combines with oxygen. These are known to cause lung problems.

3) *Sulfur dioxide* results when sulfur, which is an impurity in many fossil fuels, combines with oxygen. A pollutant in its own right, SO_2 combines with water to form sulfuric acid. This is an extremely corrosive threat to almost everything it touches.

4) *Smog* results when certain chemical reactions involving hydrocarbons take place in sunlight. Los Angeles smog is a prime example. It contains many dangerous substances such as formaldehyde and benzpyrene, a known carcinogen.

5) *Soot* refers to particulate matter that comes from engine exhaust pipes and chimneys. Aside from the dirt it produces, it is dangerous because many carcinogens stick to the tiny particles.

POLLUTION
Sources

1) *Internal combusion engines* (ICE) are the main source of carbon monoxide, hydrocarbon, and NO_x emissions. The first two occur because there is not enough air inside the cylinder of an engine. NO_x results from the high cylinder temperatures. A *diesel* engine has a higher air-to-fuel ratio than a gas engine; lower emissions of CO and hydrocarbons result unless the engine is poorly maintained. However, because combustion temperatures are higher than in a gas engine, more NO_x is produced. Also, soot is very abundant in diesel exhaust.

2) *Electrical generating plants* that use coal and oil emit most of the sulfur dioxide in the atmosphere. In addition, they are about equal to the ICE in NO_x emissions.

3) *Factories* produce the main bulk of particulates.

Effect on Climate

1) *Cities* are particularly affected by pollution because the concentration of pollutants remains high. Cities are hotter than their surroundings because the buildings absorb and retain relatively large amounts of heat. They are often referred to as *heat islands*. A thermal circulation pattern like that shown in figure 24-1 is set up. Rising air is

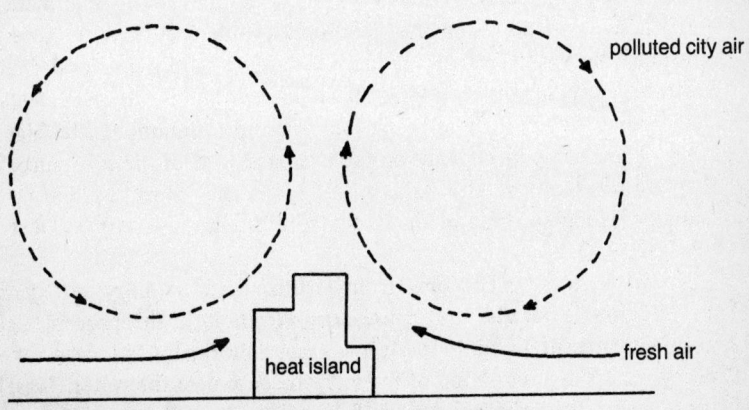

Fig. 24–1 Air circulation around heat island

replaced by cooler air from outside the city, but part of the cooler air is recycled dirty air from the city. Hence, the city is bathed in filthy air.

A *temperature inversion* makes things worse. In this situation, the temperature of the air near the ground *increases* with height (normally it decreases). This prevents an air parcel from rising because the parcel would almost immediately be cooler than its environment. The air thus stagnates and the pollution gets worse.

2) *Global climate* is affected by air pollution, but exactly how is not really known. Concentrations of CO_2 have been increasing steadily since the dawn of the industrial age. Since CO_2 absorbs heat, this should tend to increase the temperature of the earth. More evaporation should occur, and the vapor condensing on increased particulates should increase cloudiness. But this should reflect more sunlight, so the earth should cool! Whether the heating and cooling effects cancel out remains to be seen.

WATER POLLUTION

Chemical Sources

1) *Pesticides* are chemicals designed to kill particular species of living things. Aside from the obvious danger in ingesting a poison, there are two important problems associated with pesticides. First of all, most pesticides also kill species other than the intended ones. This can upset a delicate ecological balance if, say, a desirable insect is unwittingly killed by a pesticide. Secondly, some pesticides like DDT accumulate in fatty tissue. Whenever an accumulation like this occurs, the offending substance passes in ever-increasing concentrations up the food chain. Unfortunately, man is atop the food chain.

2) *Heavy metals* such as mercury, lead, and cadmium can be highly toxic. Their toxicity depends on the chemical state of the substance. In the case of the above elements, the 2^+ ion is the culprit. Mercury, for example, is highly toxic in the 2^+ state but harmless as part of a dental amalgam.

3) *Phosphates,* which come from fertilizers and laundry detergents, are undesirable because they cause *eutrophication.* In this process, algae normally present in the water grow at an abnormally fast pace. Their decay results in a reduction of the oxygen content of the water. This, in turn, makes it impossible for the fish to live. Actually, eutrophication

POLLUTION

is a natural process which normally takes thousands of years. Pollution hastens the process considerably, sometimes reducing the amount of time to less than a hundred years.

4) *Nitrates* also come from fertilizers and speed up eutrophication. In high concentrations, they are toxic to humans. They can also react chemically to form *nitrites* (NO_2^-), which are suspected carcinogens.

Nonchemical Sources

1) *Thermal pollution* results when heat produced by industry is released into a nearby body of water. This is harmful to fish because the solubility of oxygen in water decreases as the temperature rises. One way of minimizing thermal pollution is to spray the water into the air so it can cool before returning to the river or lake.

2) *Sewage* is a problem mainly because of the danger of bacteria that carry diseases such as cholera. Even if these bacteria are under control, however, the sewage can be acted upon by other microorganisms. These use up oxygen in the process, reducing the oxygen content of the water accordingly.

3) *Oil spills* float on the surface of the water because oil is less dense than water. Often, the oil floats ashore, ruining beaches and killing wildlife. Because oil does not break down readily, it stays on the surface and remains a nuisance for a long time.

Water Treatment

Typical steps in the treatment of water for a town's water supply are described below and illustrated in figure 24-2.

1) *Filtration* consists of screening out things like tin cans and dead fish.

2) During *sedimentation,* sand and other particles are allowed to settle out.

3) A *flocculating agent* is a gelatinlike precipitate which takes with it very fine particles that would otherwise remain in suspension during the sedimentation process.

4) A *sand bed* is a home for microscopic flora which consume dissolved nitrates, phosphates, and organic matter.

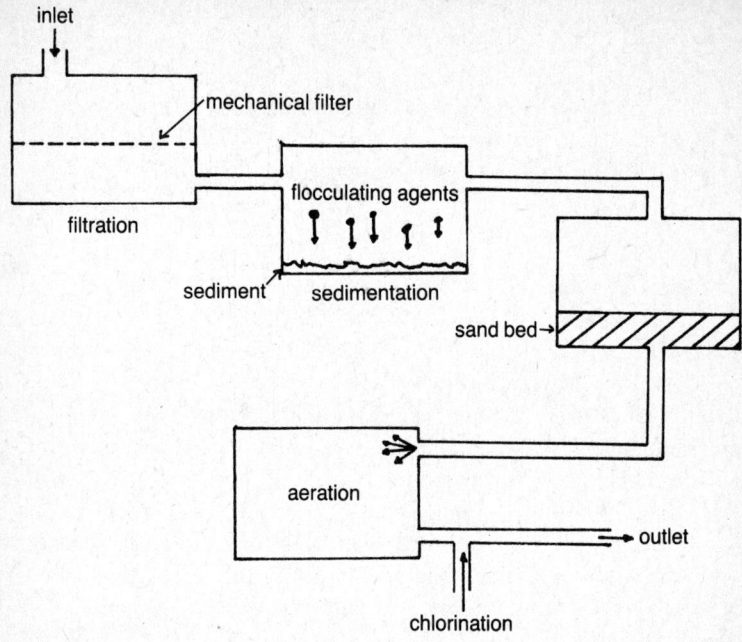

Fig. 24-2 Water treatment

5) During *aeration,* water is sprayed into the air so that it can react with oxygen. This partially purifies the water.

6) *Chlorination* is usually the final purification step. Liquefied chlorine gas is pumped into the water:

$$Cl_2 + H_2O \longrightarrow HOCl + HCl$$

HOCl is called hypochlorous acid; it is responsible for the purifying action.

There may be an undesirable side effect with chlorination. Some of the chlorine ends up reacting with various pollutants to form chloroform ($CHCl_3$) and carbon tetrachloride (CCl_4). Although the concentrations of these are small, it is possible that long-term consumption of them can cause kidney and liver problems.

In addition to the above steps, some water, called *hard water,* contains an abundance of Mg^{2+} and Ca^{2+} ions. These are not pollutants; however, they react with soap to form precipitates that build up on the inside of pipes. Water can be *softened* simply by arranging for the precipitation to occur before the water is used by the household.

Appendix

The main purpose of this appendix is to review some mathematics that could be helpful in the use of this book.

SIGNED NUMBERS

Many quantities in the physical sciences can be either positive or negative. Hence, we are often called upon to perform mathematical operations on combinations of positive and negative numbers. Since it is assumed that you can already handle positive numbers, we will discuss only cases involving negatives.

Multiplication Rules

1) The product of two negative numbers is positive.
Example: $(-2) \times (-8) = 16$
2) The product of a negative and a positive number is negative.
Example: $(-2) \times (8) = (2) \times (-8) = -16$

Division Rules

1) The quotient of two negative numbers is positive.
Example: $(-8)/(-4) = 2$
2) The quotient of a negative and a positive number is negative.
Example: $(-8)/(4) = (8)/(-4) = -(8/4) = -2$

Addition Rules

1) Two negative numbers are added by temporarily disregarding the signs, adding the numbers, and putting a negative sign in front of the result.
Example: Add $(-6) + (-8)$
Solution: Disregard signs and add: $6 + 8 = 14$
Place a negative sign in front of the result: -14

2) A negative and a positive number are added by temporarily disregarding the sign, subtracting the small number from the large one, and then placing the sign of the larger number in front of the result.
Example: Add $(-6) + (4)$
Solution: Disregard the signs and subtract: $6 - 4 = 2$
Place the sign of the larger number $(-)$ in front of the result: -2
Example: Add $(6) + (-4)$
Solution: Disregard signs and subtract: $6 - 4 = 2$
Place the sign of the larger number $(+)$ in front of the result: $+2$

Subtraction

There is only one rule for subtraction: change the sign of the number being subtracted, then perform *addition*, not subtraction.
Example: Subtract $(4) - (8)$
Solution: Change the sign of the number being subtracted: $+8 \rightarrow -8$
Perform the *addition:* $(4) + (-8) = -4$
Example: Subtract $(-8) - (-4)$
Solution: Change the sign of the number being subtracted: $-4 \rightarrow 4$
Perform the *addition:* $(-8) + (4) = -4$

EXPONENTS

Sometimes a quantity has to be multiplied by itself many times over. There is a simple way of doing this:

$$x \cdot x \cdot x \cdot x = x^4$$

Here, x is called the *base*, and the 4 is called the *exponent*. The notation x^4 is read "x to the fourth power." There are two simple rules that cover multiplication and division of numbers with the same base but different exponents.
Rule 1: $x^a \cdot x^b = x^{a+b}$
Example: $10^3 \cdot 10^2 = 10^{3+2} = 10^5$

APPENDIX

Rule 2: $\dfrac{x^a}{x^b} = x^{a-b}$

Example: $\dfrac{10^4}{10^2} = 10^{4-2} = 10^2$

Exponents can also denote more than just repeated multiplication. For instance:

Definition: $x^{-a} = \dfrac{1}{x^a}$

Example: $10^{-2} = \dfrac{1}{10^2} = .01$

Also, $x^0 = 1$. This follows from the division rule. We know that $\dfrac{x^a}{x^a} = 1$ since any number divided by itself is 1. But $\dfrac{x^a}{x^a} = x^{a-a}$, which is x^0. Hence, $x^0 = 1$.

Finally, we should note that several rules can be used in one problem:

Example: $\dfrac{x^3 \cdot x^{-8}}{x^{-5}} = \dfrac{x^{3-8}}{x^{-5}} = \dfrac{x^{-5}}{x^{-5}} = x^{-5-(-5)} = x^0 = 1$

POWERS OF TEN AND SCIENTIFIC NOTATION

By *powers of ten* we mean 10^i, where i is an integer. We are especially interested in multiplying and dividing by powers of ten. Doing so involves only the moving of a decimal point a number of places equal to the value of the exponent. Two rules cover whether the movement is left or right.

Rule 1: The decimal point is moved i places to the right when multiplying by 10^i or dividing by 10^{-i}.

Example: $3.574 \times 10^4 = 35740$ (moved 4 places to right)

Example: $3.574 \div 10^{-2} = 357.4$ (moved 2 places to right)

Rule 2: The decimal point is moved i places to the left when multiplying by 10^{-i} or dividing by 10^i.

Example: $3.574 \times 10^{-3} = .003574$ (moved 3 places left)

Example: $3.574 \div 10^1 = .3574$ (moved 1 place left)

Scientific notation is a way of expressing a particular number such that another number between one and ten, multiplied by a power of ten, is equal to the original number. Two examples will illustrate:

$$3,500,000,000,000 = 3.5 \times 10^{12}$$
$$.00000000035 = 3.5 \times 10^{-10}$$

In each case the 3.5 is called the *mantissa*.

It is important to know how to put numbers in scientific notation and how to use those numbers in calculations.

Putting a Number in Scientific Notation

This requires a few simple steps:
1) With the digits in their original order, put the decimal point in the place which turns the digits into a number between one and ten.
2) Count how many places you must move the decimal point in order to bring it from the new position to the original position. Call this number n. It is positive if you moved rightward to get back to the original position, negative if you moved leftward.
3) Multiply the number in 1) by 10^n.

Looking at our examples, we see that in both cases the decimal point was moved between the three and the five in order to get a number between one and ten. In the first one, we had to move twelve places to the right in order to get the decimal point back to the original place. Hence we multiplied by 10^{12}. In the second, we moved ten places to the left; hence the multiplier was 10^{-10}.

Calculations with Scientific Notation

These usually involve only multiplication and/or division.

Example: Calculate $\dfrac{(4.91 \times 10^8) \times (6.18 \times 10^4)}{2.81 \times 10^{-3}}$

Solution: Rewrite as

$$\left(\frac{4.91 \times 6.18}{2.81}\right) \times \left(\frac{10^8 \times 10^4}{10^{-3}}\right)$$

Note that the powers of ten are grouped together.

We can use a simple calculator for the first part:

$$\frac{4.91 \times 6.18}{2.81} = 10.8$$

We use the rules about powers of ten to do the second part:

$$\frac{10^8 \times 10^4}{10^{-3}} = \frac{10^{12}}{10^{-3}} = 10^{12-(-3)} = 10^{15}$$

The final answer is then:

$$10.8 \times 10^{15} \quad \text{or} \quad 1.08 \times 10^{16}$$

Scientific notation makes the problem easy because the multiplication and division is done with numbers that are between one and ten. The powers of ten are treated by simple addition or subtraction.

APPENDIX

MEASUREMENT

There are three fundamental quantities upon which all other quantities in the physical sciences are based: length, mass, and time. The meaning of these terms seems evident to all of us, yet defining them in the dictionary sense is quite unsatisfactory. (Try looking them up!) Fortunately, such definitions are not needed. What is important is that we can *measure* them.

Every measurement has two parts: a numerical part and a units part. The *units* identify what sort of quantity we are talking about; the *numerical part* tells us how much we have. Thus, the length of a ruler is not 12, but 12 *inches*.

Metric System

The *metric system* is a set of standardized units used in the measurement of physical quantities.

1) The *meter* (m) is the standard unit of length. This is about the length of your arm.

2) The *kilogram* (kg) is the standard unit of mass. This is approximately the mass of a quart of milk.

3) The *second* is the standard unit of time; we are all familiar with this unit.

All other quantities have units which are a particular combination of the standard ones. For example, consider the area of a rectangular surface, which is defined as the product of the two perpendicular lengths. A rectangle measuring 2 m by 3 m has an area of 2 m × 3 m, according to our definition. Mathematically, this is $2 \times 3 \times m \times m$ or 6 m^2. Note that the units in a formula obey the same algebraic rules as the numbers. Our result, 6 m^2, means that the units of area are meters squared and that this surface has an area equal to 6 of those units. Similarly, the *volume* of a box is the product of three perpendicular lengths. A box measuring 1 m by 2 m by 4 m has a volume of 8 m^3, using the above reasoning. The units of volume are meters cubed and this box has a volume equal to 8 of those units.

The system that uses the meter, kilogram, and second as the basic units is called the *SI system*. It was agreed to standardize things this way only about twenty years ago. Hence, there are other systems still in use. For example, there is another system involving metric units which has the centimeter, gram, and second as its basic units. This is called the *cgs system;* when appropriate, we have given certain quantities in their cgs units.

Conversion of Units

Conversion of units means changing a quantity measured in a particular set of units to the same quantity measured in different units. It is sometimes inconvenient to express certain quantities in the standard units. For instance, it is clumsy to use meters to measure the thickness of a penny. When a situation like this arises, a new unit can be defined which is a multiple or fraction of the old unit. The new unit is defined in terms of a conversion equation, several of which are given in Table A-1.

TABLE A-1. CONVERSION EQUATIONS

1 centimeter (cm) = .39 inch (in)
1 meter (m) = 100 cm = 1000 millimeters (mm)
1 liter (l) = 1.06 quart (qt) = 1000 milliliters (ml)
1 kilogram (kg) = 1000 grams (g)
1 g = 1000 milligrams (mg)
1 kg = 2.2 pounds (lb)

This table is used in conjunction with the following rule:

Conversion Rule: Find an equation that relates the new and old units. Multiply the original quantity by a fraction whose numerator is the side of the equation with the new units and whose denominator is the side with the old units.

Example: Express 3.5 m in mm.

Solution: The new units are millimeters (mm) and the old ones are meters (m). Our conversion equation is 1 m = 1000 mm. We multiply the original quantity by 1000 mm/1 m, i.e., new units over old units:

$$3.5 \text{ m} \times \frac{1000 \text{ mm}}{1 \text{ m}} = 3500 \text{ mm}$$

Example: Express 2.2 inches (in) in centimeters (cm).

Solution: The new units are centimeters, the old ones are inches. The conversion equation is 1 cm = .39 in. Using the rule we get:

$$2.2 \text{ in} \times \frac{1 \text{ cm}}{0.39 \text{ in}} = 5.6 \text{ cm}$$

In the first example, we ended up multiplying by 1000. In the second one we ended up dividing by 0.39. It is important to realize that given a table of conversions, we never have to memorize whether to multiply or divide in a given situation. When properly used, the conversion rule automatically sets up the correct arithmetic operation.

APPENDIX

Continuous and Discrete Quantities

1) A *continuous quantity* is one which has an infinite number of values between two measurements of that quantity. Your height, for example, is a continuous quantity because between any two heights you can have an infinite number of different heights.

2) A *discrete quantity* is one which has only a finite number of values between two measurements. The number of chairs in a room, for example, is a discrete quantity. Between the measurements 30 chairs and 35 chairs you can have only four other measurements: 31, 32, 33, and 34. All other values are impossible; for instance, what does 34.5 chairs mean?

As new quantities are introduced in the chapters on physics, you will undoubtedly conclude that they are continuous. This is exactly the way physicists viewed them until roughly the turn of the century. One of the most remarkable discoveries in physics was the revelation that quantities which appeared continuous actually turned out to be discrete. This formed the basis for a revolution in physical thought.

Significant Figures

Significant figures are particular digits that are used to record the measurements of physical quantities. Assuming that a measurement does not turn out to be an integer (this is almost always the case), we can use a simple two-step process to determine the significant figures.

Step 1: Record the measurement using all digits that are certain, plus one more digit to the right of the decimal point.

Example: How should the length of the rods in figure A-1 be recorded?

Solution: a) This ruler is marked with lines spaced every 1 cm. We record the length of the ruler as 4.3 cm. We are certain of the digit 4. However, we can only estimate the digit 2 by noting that the end of the rod is about three tenths of the distance from the 4 to the 5.

b) This ruler is marked with lines every 0.1 cm. We record the length as 4.25 cm. We are certain of the 4 and the 2. However, we can only estimate the 5 by noting that the end of the rod is about halfway between the 4 and the 5.

c) In this case, the end of the rod appears to be just at the 4.3 cm mark. We must record the measurement as 4.30 cm, not 4.3 cm. The latter result implies that we are sure of the 4 but have estimated the 3. In fact, we are sure of both the 4 *and* the 3. That is why the 0 is part of this measurement.

Step 2: Express the measurement using scientific notation, keeping all digits *except* zeros that end up to the left of the decimal point.

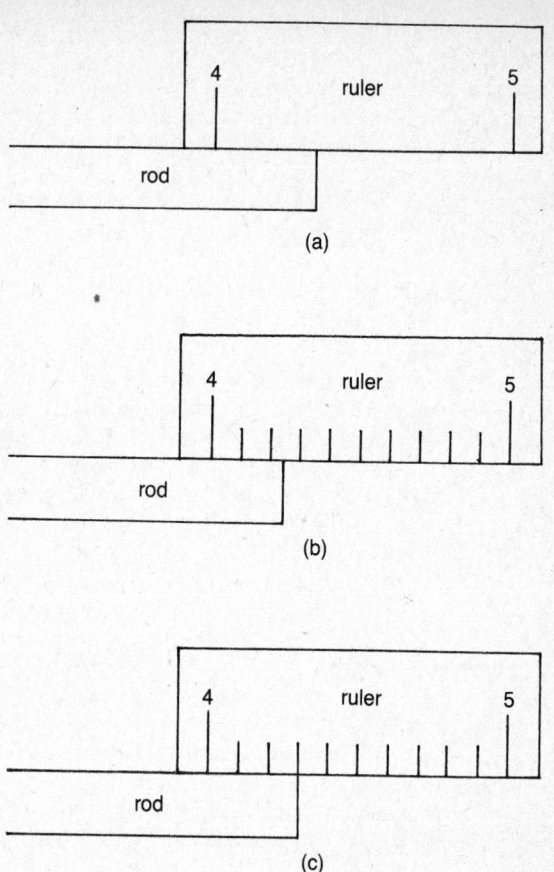

Fig. A–1 Measurement of a rod

Example: How would the following measurements appear in scientific notation?
a) 27.4 cm   b) 0.00025 kg   c) 27.400 kg
Solution: a) 2.74×10^1 cm

b) 2.5×10^{-4} kg. Note that after the decimal point has been moved to between the 2 and the 5, all the zeros to the left are dropped.

c) 2.7400×10^1 kg. In this case, we didn't drop the zeros. They are not to the left of the decimal point.

Once the measurement is correctly recorded and put into scientific notation, finding the significant digits is easy: *The significant digits are those in the mantissa.*

APPENDIX

Example: How many significant digits are there in the measurements of the previous example?
Solution: a) three—2, 7, 4
  b) two—2, 5
  c) five—2, 7, 4, 0, 0

1) For *measurements that are integers*, the only problem is what to do about zeros at the end of the number. Unless otherwise noted, it is assumed that zeros at the end of an integer are not significant.

Example: An archaeologist determines that a bone is a million and a half years old. How many significant figures are there in this measurement?
Solution: Since the zeros to the right are not significant, we will drop them after the number is put into scientific notation. Hence, 1,500,000 is written as 1.5×10^6, not 1.500000×10^6. Hence, there are two significant figures.

2) When *multiplying or dividing measurements*, the result must be rounded off to the number of significant figures found in the measurement with the least number of significant figures.

Example: Multiply 3.042 cm by 2.5 cm.
Solution: Using a calculator, we get 7.605 cm^2 as the answer. However, we must round off to two significant figures because one of the measurements has only two significant figures. Hence, the answer is recorded as 7.6 cm^2.

3) When *adding or subtracting measurements*, the result must be rounded off to the same number of places to the right of the decimal point as the measurement with the least number of places to the right of the decimal point.

Example: Add 26.042 g + 7.38 g + 127.9 g.
Solution: Mathematically, the result would look like this:

$$\begin{array}{r} 26.042 \text{ g} \\ 7.38 \text{ g} \\ \underline{127.9 \text{ g}} \\ 161.322 \text{ g} \end{array}$$

This must be rounded to one place because the measurement 127.9 has the fewest number of places, namely one. Hence, the answer is recorded as 161.3 g.

4) Numbers which are *not measurements* are assumed to have an infinite number of significant digits.

Example: What is the result of multiplying 3.105 cm by 2?
Solution: If the 2 were a measurement, say, 2 cm, it would have only one significant digit. Then the answer, 6.210 cm^2, would have to be rounded to one significant digit, 6. However, the number 2 is *not* a measurement. Then the number of significant digits in the answer is

limited only by the number in the measurement. Hence, the answer retains its four significant digits.

SIMPLE EQUATIONS

The expressions

$$x + 3 = 10$$
$$x - 8 = 5$$
$$2x = 12$$
and
$$\frac{x}{3} = 5$$

are true statements only if an appropriate value of x is used in each case. Finding the correct value of x for an equation is called *solving* the equation for the *unknown*, x.

Basic rule for solving equations: Perform any correct mathematical step(s) to both sides of the equation so that the unknown is isolated on one side.

Example: $x + 3 = 10$
Solution: $x + 3 \underline{-3} = 10 \underline{-3}$ (subtract 3 from both sides)
$x = 7$ (unknown isolated on one side)

Note that we could perform *any* correct operation on both sides. For instance, we could add 5:

$$x + 3 \underline{+5} = 10 \underline{+5} \text{ (add 5 to both sides)}$$
$$x + 8 = 15$$

While this is a true statement, the unknown is still not isolated. Obviously, we must perform not only a correct step, but also a reasonably clever one.

Example: $x - 8 = 5$
Solution: $x - 8 \underline{+8} = 5 \underline{+8}$ (add 8 to both sides)
$x = 13$ (unknown isolated)

Example: $2x = 12$
Solution: $\frac{2x}{2} = \frac{12}{2}$ (divide both sides by 2)
$x = 6$ (unknown isolated)

Example: $\frac{x}{3} = 5$
Solution: $\frac{3 \cdot x}{3} = 3 \cdot 5$ (multiply both sides by 3)
$x = 15$ (unknown isolated)

APPENDIX

Sometimes the unknown is in the denominator. This presents no special problem.

Example: Solve for x: $\dfrac{8}{x} = 2$

Solution: Multiply both sides by x:

$$x \cdot \frac{8}{x} = 2x$$

The resulting equation is $8 = 2x$
Divide by 2 to solve: $x = 4$

FORMULA EVALUATION

A *formula* is an equation that shows a relationship between various quantities. Evaluating a formula consists of substituting numerical values for all but one of the quantities, then solving the resulting equation for the remaining quantity, which is the unknown.

Example: Suppose a law is represented by the equation $A = \dfrac{BC}{D}$

Find C if $A = 10$, $B = 8$, and $D = 4$.

Solution: Substitution of these numbers gives the following equation:

$$10 = \frac{8C}{4} = 2C$$

Solving the resulting equation for C, we get $C = 5$.

ALGEBRAIC FRACTIONS

Manipulation of formulas and equations often requires multiplication and division of fractions.

1) The *product of two fractions* is a fraction whose numerator is the product of the original numerators and whose denominator is the product of the original denominators.

Example: What is $\dfrac{a^2}{c} \times \dfrac{b}{a}$

Solution: The new numerator is $a^2 b$
The new denominator is ca
The answer is $\dfrac{a^2 b}{ca} = \dfrac{ab}{c}$

Note that a factor a cancels out. Always reduce to lowest terms.

2) To *divide two fractions*, invert the divisor and multiply.

Example: What is $\dfrac{a^2}{c} \div \dfrac{b}{c}$

Solution: Invert the divisor $\dfrac{b}{c}$ and multiply: $\dfrac{a^2}{c} \times \dfrac{c}{b} = \dfrac{a^2 c}{cb}$

After canceling, we get $\dfrac{a^2}{b}$

3) A *compound fraction* is one fraction over another. This often appears in physical problems. It is nothing more than a problem of division of fractions.

Example: What is $\dfrac{\dfrac{b^2 c^2}{a}}{\dfrac{c}{a}}$

Solution: This is a division problem:

$$\frac{b^2 c^2}{a} \div \frac{c}{a} = \frac{b^2 c^2}{a} \times \frac{a}{c} = \frac{b^2 c^2 a}{ac} = b^2 c$$

GRAPH INTERPRETATION

A *graph* is a picture which shows how one physical quantity depends on another. Figure A-2 is an example of a graph. The two solid perpendicular lines are called the *coordinate axes*. The horizontal one is called the *x-axis;* the vertical one is called the *y-axis*. Their intersection is called the *origin*. A wide range of possible values of each one of the quantities is marked off on its own axis as shown.

The graph enables us to do two important things. First, we can get a quick summary of the general behavior of the quantities. In this case, when one quantity increases, the other decreases. Second, if we know a particular value of one quantity, we can draw perpendicular lines to the graph, as shown, and read off the value of the other. In this instance, we can say that $y = 1$ if we know that $x = 4$.

SCALARS AND VECTORS

A *scalar* is a quantity having only magnitude. A *vector* is a quantity having both magnitude and direction. A vector is symbolized by a letter in boldface type, such as **V**. Mass, for example, is a scalar. There is no direction associated with it. Force, on the other hand, is a vector. The direction in which you push or pull something is just as important as how hard you exert yourself.

APPENDIX

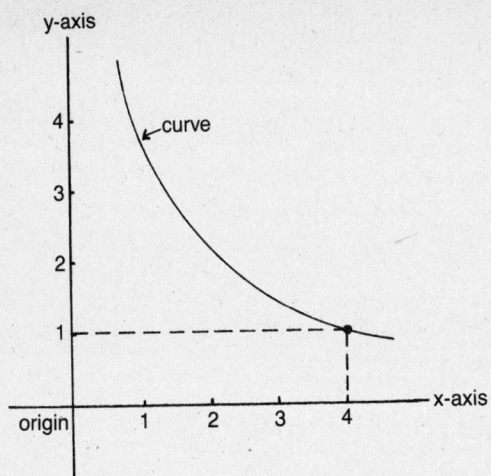

Fig. A-2 A typical graph

Representation of a Vector

A vector is represented by an arrow whose length is proportional to the magnitude and whose direction is the direction of the vector. The front is called the *head* and the rear is called the *tail*. Any two arrows with the same length and direction represent the same vector, even though they are not in the same place. Thus, an arrow representing a vector can be moved about. As long as it has the same direction and length, it is the same vector.

The negative of a vector **V** is another vector which has the same magnitude but points in the opposite direction. We call the new vector −**V**.

Addition and Subtraction of Vectors

1) To *add* two vectors, keep either one stationary and move the other so that the tail of the moved one is on the head of the stationary one. The sum of these two, called the *resultant*, is a vector drawn from the tail of the stationary one to the head of the moved one. This is illustrated in figure A-3.

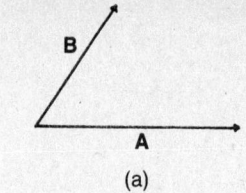

(a)

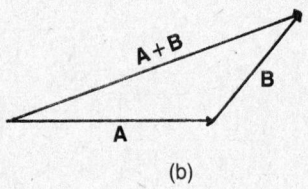

(b)

Fig. A–3 Addition of two vectors

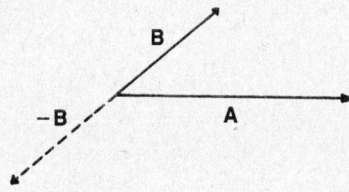

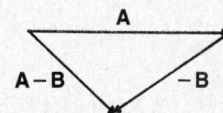

Fig. A–4 Subtraction of two vectors. $\mathbf{A} - \mathbf{B} = \mathbf{A} + (-\mathbf{B})$.

APPENDIX

2) To *subtract* two vectors, say **A** minus **B**, reverse the direction of **B**, making it −**B**, then add up **A** and −**B**. This is illustrated in figure A-4.

3) A vector can be *multiplied or divided* by a scalar. The *direction* is unchanged; the *magnitude* gets multiplied or divided accordingly.

Resolution of Vectors

Resolving a vector means finding two other vectors, called *components*, that add up to a given vector. There are an infinite number of pairs of components that add to the given vector. To avoid confusion, it is conventional to choose the case whose components are parallel to the coordinate axes. The *x-component* is the one that is parallel to the x-axis; the *y-component* is parallel to the *y*-axis. Figure A-5 shows the resolution of the vector V into its components.

Very often, only one component of a vector is of interest in a given problem. For example, say that V in figure A-5 represents the pull of a child on a sled. The motion of the sled to the right depends on $\mathbf{V}_x$, not on $\mathbf{V}_y$.

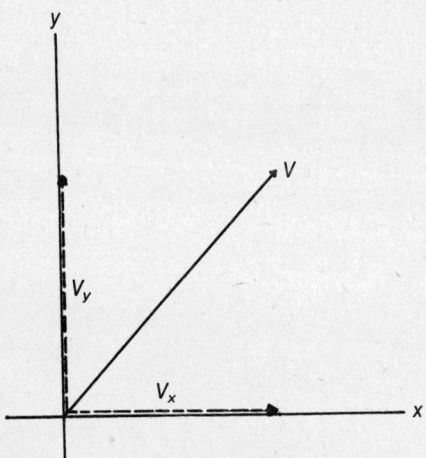

Fig. A–5 Resolution of a vector into *x*- and *y*-components.

DIRECT AND INVERSE PROPORTION

Direct Proportion

Two quantities are said to be *directly proportional* if an increase (decrease) in one is matched by a proportionate increase (decrease) in the other. If x and y are variables which represent the two quantities, then we can write:

$$y = kx$$

Here, k is any fixed number; it is called a *constant of proportionality*. In different problems, it takes on different values. In a given problem, however, it remains unchanged. Its value does not affect the *behavior* of the relation, as we can see by comparing the following examples.

Example: Given $y = 4x$. What happens to the value of y when x goes from 8 to 10, an increase of 1/4 the original value?
Solution: When $x = 8$, $y = 4 \times 8 = 32$
When $x = 10$, $y = 4 \times 10 = 40$
Hence, y increases from 32 to 40, an increase of 1/4 its original value. In other words, y changes exactly the same way as x.

Example: Given $y = 3x$. What happens to y under the same conditions as above?
Solution: When $x = 8$, $y = 3 \times 8 = 24$
When $x = 10$, $y = 3 \times 10 = 30$
Hence, y increases from 24 to 30, also an increase of 1/4 its original value. From this we see that y and x change the same way, regardless of the value of k.

Inverse Proportion

Two quantities are said to be *inversely proportional* if one increases while the other decreases by a proportionate amount. This is represented by the equation:

$$y = k/x$$

Example: Given $y = 20/x$. What happens to y when x goes from 5 to 10?
Solution: When $x = 5$, $y = 20/5 = 4$
When $x = 10$, $y = 20/10 = 2$
We can see that as x doubles, y gets cut in half.

APPENDIX

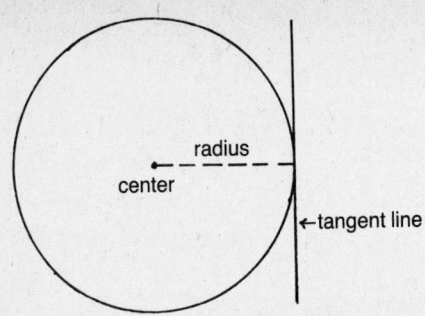

Fig. A-6 The circle. The tangent and radius are perpendicular.

SOME FACTS ABOUT CIRCLES, PARABOLAS, AND ELLIPSES

Circles

For a circle of radius R:

1) The *circumference is* $2\pi R$.
2) The *area* is πR^2.
3) A *tangent line* is a line that touches the circle at one point and is perpendicular to a radius drawn to that point (figure A-6).

Parabolas

A *parabola* is a curve (see figure A-7) that is related to a fixed point, F, and a fixed line, D, in the following way: any point, P, on the parabola is equidistant from F and D.

Projectiles such as baseballs and artillery shells sail through the air in a parabolic path.

Ellipses

An *ellipse* is a curve (see figure A-8) with the following relation to two fixed points, F_1 and F_2: no matter where P is, the sum of the distances $F_1 P + F_2 P$ is constant. For reference, F_1 and F_2 are called *foci*, AB is called the *semimajor axis*, and AC is called the *semiminor axis*.

The path of any planet is an ellipse.

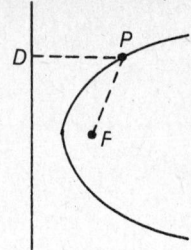

Fig. A-7 The parabola. *PD* always equals *PF*.

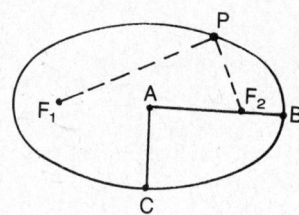

Fig. A-8 The ellipse. $F_1P + F_2P$ always add up to the same value.

TEMPERATURE CONVERSIONS

1) Celsius to Fahrenheit: $F = 1.8C + 32$

Example: Convert 32°C to Fahrenheit.
Solution: Multiply the 32 by 1.8: $32 \times 1.8 = 57.6$
Add 32: $57.6 + 32 = 89.6$
Thus, 32°C = 89.6°F

2) Fahrenheit to Celsius: $C = \dfrac{5(F - 32)}{9}$

Example: Convert 60.5°F to Celsius.
Solution: First subtract 32 from 60.5: $60.5 - 32 = 28.5$
Multiply the result by 5: $5 \times 28.5 = 142.5$
Divide the above by 9: $142.5 \div 9 = 15.8$
Thus, 60.5°F = 15.8°C

3) Celsius to Kelvin: $K = C + 273$

Example: 25°C = 25 + 273 = 298°K

4) Kelvin to Celsius: $C = K - 273$

Example: 400°K = 400 - 273 = 127°C

5) To convert between Fahrenheit and Kelvin, convert first to Celsius and use the above rules.

Index

Absolute dating, 277
Absolute magnitude, 199
Abyssal plain, 253
AC generator, 61
Acceleration, 5, 6
  centripetal, 17
  gravitational, 10
Acid, 135–136, 152
Action reaction pair, 8
Activation energy, 129
Activity series, 141
Addition reaction, 152
Aeration, 286
Aesthenosphere, 273
Air mass, 237
Airplane wing, 38
Alcohols, 151
Aldehydes, 151
Alkali metals, 116
Alkaline earth, 116
Alkanes, 149
Alkene, 150
Alkyl group, 150
Alkynes, 150
Alluvium, 250
Alpha decay, 164
Alpha particle, 162
Amide, 152
Amines, 152
Amino acid, 156
Ampere, 53, 56
Amplitude, 70
Angular momentum, 22
  speed, 14, 15
  velocity, 15
Anode, 143
Anticlines, 267

Anticyclonic flow, 223
Apparent magnitude, 199
Aquifer, 270
Arch, 256
Archimedes' principle, 34, 269
Aromatic hydrocarbons, 150
Asteroid, 180
Atmospheric pressure, 33
Atom, 103
  and spectra, 107
  Bohr model, 107
  quantum mechanical model, 109
  Rutherford model, 104
  Thompson's model, 104
Atomic mass (amu), 106
  number, 106
Aurora, 180
Autoionization, 135
Automobile brake system, 37

Barometer, 33
Barycenter, 184
Base, 136, 157, 288
Base level, 250
Batholith, 266
Battery, 53
Bergeron-Findeisen process, 236
Bernoulli's principle, 37
Beta decay, 165
  particle, 163
Big Bang theory, 207
Binary star, 202
Binding energy, 166
Black dwarf, 203
Blackbody, 90
Blue shift, 202
Bode's law, 197

INDEX

Boiling, 46
Bond, covalent, 119–120
  double, 119
  ionic, 117
  triple, 120
Boyle's law, 45
Brownian movement, 39
Buoyancy, 34, 213

Caldera, 266
Calorie, 40
Capacity, 250
Carbohydrate, 154
Carbon dioxide, 212
Carbon monoxide, 282
Carboxylic acid, 152
Catalyst, 130
Cathode, 143
  rays, 103
Cavern, 270
Celestial coordinates, 192
  sphere, 191
Cell, Daniell, 144
  lead storage, 146
  membrane, 158
  voltaic, 144
Cellulose, 154
Cenozoic era, 281
Centigrade (Celsius) scale, 40
Centrifugal force, 18–20, 223
Centripetal acceleration, 17
  force, 18
Chain reaction, 168
Channel shape, 250
Charge, 50
  of electron, 104
  of proton, 104
Charle's law, 45
Chemical reaction:
  addition, 154
  combination, 126
  combustion, 127
  condensation, 154
  decomposition, 127
  displacement, 127
  exchange, 127
  polymerization, 153
  substitution, 153
Chlorination, 286
Chromosomes, 158
Chromosphere, 179
Circle, 303
Circular motion, 13–20

Circulation index, 230
Cirque, 256
Cleavage, 247
Climate, 211
Cloud chamber, 164
Clouds, 234, 241
Coalescence, 236
Coefficients, 125
Coinage metals, 116
Color index, 200
Coma, 197
Combination reaction, 126
Combustion reaction, 127
Comets, 197
Commutator, 61
Competence, 250
Compound, 117
  covalent, 119
  ionic, 117–118
  organic, 148–156
  stability, 130
Condensation, 31
  nucleii, 234
  reaction, 154
Conduction (of heat), 42
Conductor (electric), 53
Conservation:
  of energy, 27
  of mass, 125
  of momentum, 21
Constellation, 192
Continental margin, 253
  rise, 253
  shelf, 253
  slope, 253
Continuous quantity, 293
Control rods, 168
Convection, 42
Convergence, 223
Coordinate axes, 298
Core, 247
Coriolis effect, 221
Corona, 179
Corpuscular theory, 78
Correlation, 279
Cosmic rays, 206
Cosmology, 206
Coulomb, 51
Coulomb's law, 51
Crater, 188, 266
Creep, 258
Crest, 70
Critical mass, 168

INDEX

Crosscutting, 279
Crust, 247
Curie point, 274
Current, 53, 56
   AC, 60–66
   DC, 53–56
   and magnetic field, 60
Curvature effect, 235
Cutoff frequency, 92
Cyclic theory, 207
Cyclogenesis, 240
Cyclone, 237
Cyclonic flow, 223

Daniell cell, 144
Day, 173
Declination, 192
Decomposition reaction, 127
Deflation, 257
Delta, 250
Density, 31, 216
Desert pavement, 257
Dew, 237
Dew point, 237
Diffraction, 76
Dike, 266
Discacharides, 154
Discharge, 250
Discrete quantity, 293
Displacement, 3
   angular, 4
   reaction, 127
Dissociation, 133
Distance, 3
Divergence, 223
Divide, 251
DNA, 157
Doppler effect, 77
Drainage basin, 251
Dry atmosphere, 211
Dry cell, 147
Dwarf stars, 202, 203
Dynamic lifting, 234

Earthflow, 259
Earthquakes, 260–263
Easterly waves, 243
Ecliptic, 192
Ekman spiral, 254
Elastic rebound, 260
   strain, 260
Electric:
   current, 53–56

   field, 51, 52, 56
   motor, 61
Electrolysis, 142
Electrolyte, 133
Electrolytic cell, 142
Electromagnet, 60
Electron, 53, 104
   configuration, 113
   spin, 112
Electroplating, 143
Element, 114
   electronic configuration, 113
   families, 116, 123
   periodic table, *front cover,* 114
   stability, 116
Ellipse, 303
Endothermic reaction, 129
Energy, 24–29, 39
   activation, 129
   binding, 166
   chemical, 128–130
   conservation, 27
   electronic, 103–104
   kinetic, 25
   level, 107
   potential, 25
   transformation, 27
Enzyme, 130
Epicenter, 261
Epoch, 280
Equation
   algebraic, 296
   chemical, 125, 126, 127, 129, 130
   nuclear, 164–165
Equator, 171
Equatorial coordinate system, 191
   convergence zone, 229
Equilibrium:
   chemical, 137
   of air parcel, 232–233
   thermal, 40
Equinox, 176
Era, 280
Erosion, 254–257
Ester, 152
Ether, 78
Eutrophication, 284
Evaporation, 31, 46
Exchange reaction, 127
Exfoliation, 258
Exothermic reaction, 129
Exponent, 288
Extended source, 72

Eye (hurricane), 243
Eyeglasses, 86
Eyepiece, 89

Farsighted eye, 86
Fats, 155
Faults, 260–261
Ferromagnetism, 57
Filament, 180
Filtration, 285
Fiord, 256
Fission, 167
Fissure, 264
Flocculating agent, 285
Floodplain, 251
Fluid, 30
Focal:
　length, 83
　point, 83
Focus, 261, 303
Fog, 237
Folds, 267
Force, 7
　centrifugal, 18–20, 223
　centripetal, 18
　Coriolis, 221
　electric, 51
　gravitational, 9
　lines of, 52
　magnetic, 56
　nuclear, 165
　pressure gradient, 220
Formula, 117, 121–122, 148, 297
Fossils, 275, 279, 281
Frame of reference, 3, 94–99
Free fall, 9
Freezing, 31
　nucleii, 234
Frequency, 70
Friction, 11, 223, 225
Fronts, 237–238
Frost, 237
　wedging, 258
Fructose, 154
Functional group, 149
Fundamental, 71
Fusion, 166

Galaxy, 204
Galvanometer, 64
Gamma ray, 163

Gas, 31
　constant, 138
　ideal, 44–46
　laws, 45
　noble, 116
Geiger counter, 163
Gene, 158
General circulation, 228
Genetic code, 157
Geocentric theory, 189
Geostrophic wind, 224
Geosyncline, 277
Geyser, 272
Giant star, 202
Glacier, 252
Glucose, 154
Glycogen, 154
Gondwanaland, 273
Graben, 268
Gradient (of stream), 250
Gradient wind, 225
Granule, 179
Graph, 298
Gravitational:
　force, 9
　mass, 98
　potential energy, 25
Great circle, 171
Groundwater, 270
Group, 116
　functional, 149
Guyot, 253

Hadley cell, 231
Hail, 236
Half-life, 163
Half-reaction, 140
Halogens, 116
Hard water, 286
Hardness, 248
Harmonic, 71
Heat, 40
　engine, 47
　of fusion, 44
　island, 283
　latent, 44, 213
　reservoir, 47
　transfer, 42, 49
　of vaporization, 44
Heliocentric theory, 189
Hertzsprung-Russell diagram, 201
Horn, 256

INDEX

Horse latitudes, 229
Horsepower, 24
Horst, 268
Hubble's law, 205
Hurricane, 243
Hydration, 135
Hydraulic lift, 36
Hydrocarbons, 149
Hydronium, 135

Ice Age, 281
Ideal gas, 44–46
Ideal gas law, 45
Igneous rocks, 248
Inertial mass, 98
Inner transition elements, 117
Insulator, 53
Interference, 75–76
Internal energy, 39
International dateline, 175
Ion, 106, 134–137, 140–147
Ionic equation, 134
Ionosphere, 215
Isobar, 220
Isomer, 149
Isostasy, 268–269
Isotope, 106

Jet stream, 230
Jointing, 258
Joule, 23
Jupiter, 195

Kepler's laws, 190
Ketones, 152
Kettle, 252
Kilogram, 291
Kinetic energy, 25
  translational, 39

Laccolith, 266
Lake, 270
Lapse rates, 232–233
Latent heat, 44, 213
Laterite, 249
Latitude, 173
Laurasia, 273
Lava, 264
Le Châtelier's principle, 138
Lead storage cell, 146
Length contraction, 96
Lenses, 83

Levee, 251
Light, 78–81
  diffraction, 76
  interference, 75–76
  polarization, 81
  reflection, 72
  refraction, 74
  scattering, 79
  speed, 95
  theories, 78
  year, 201
Lightning, 242
Linear motion, 4
  vs. circular motion, 15
Lines of force, 52
Liquids, 31
Load, 250
Longitude, 173
Lorentz transformation factor, 99
Lunar eclipse, 182

Magnetic:
  domain, 56
  field, 56, 58, 59
  moment, 56
  poles, 57
Magnifying glass, 86
Magnitude (of star), 199
Main sequence, 201
Mantissa, 289
Mantle, 247
Maria, 188
Mars, 194
Mascon, 188
Mass:
  atomic, 106
  conservation, 125
  and energy, 97
  gravitational, 98
  inertial, 98
  unit, 291
  wasting, 258
Meander, 251
Mechanical lifting, 234
Meiosis, 159
Melting, 31
Meltwater, 252
Mercalli scale, 261
Mercury, 194
Meridian, 173
Meridional flow, 228
Meson, 165

Mesopause, 214
Mesosphere, 214
Mesozoic era, 281
Metals (vs. nonmetals), 114
Metamorphic rocks, 249
Meteor shower, 198
Meteorite, 198
Meteoroid, 198
Meter, 291
Michaelson-Morley experiment, 95
Micrometeorite, 198
Milky Way, 205
Millibar, 216
Mineral, 247
   examples, 248
Mirrors, 83
Mitosis, 159
Mixing ratio, 213
Molarity, 133
Mole, 128
Molecular weight, 120
Molecule, 120
Moment of inertia, 17
Momentum, 21-22
Monomer, 153
Monosaccharides, 154
Moon:
   gravitation, 10
   orbit, 181, 184
   phases, 181
   surface, 186
   and tides, 183-186
Moraines, 252
Mountain, 267-269
Mudflow, 259

Nearsighted eye, 86
Neptune, 196
Net ionic equation, 135
Neutron, 104, 106
   star, 203
Newton's laws:
   first, 7
   gravitational, 9
   second, 8
   third, 8
Nitrates, 285
Nitrites, 285
Nitrogen, 211
   oxides, 282
Noble gases, 116
Nuclear equation, 164
   force, 165

Nucleic acid, 157
Nucleotide, 157
Nucleus, 104, 158, 162, 197
Nuee ardente, 264

Objective, 86
Oblique ray, 176
Oceans, 253-254
Ohm, 53
Ohm's law, 54
Oil, 155
Orbitals, 110, 111
   designation, 111
   energies, 112
Origin (of coordinate system), 298
Outwash plain, 252
Oxbow lake, 251
Oxidation, 140
Oxygen, 211
Ozone, 212

Paleomagnetism, 274
Paleozoic era, 280
Pangaea, 273
Parabola, 83, 303
Parallel (on earth), 171
Parallel circuit, 55
Parcel, 220
Parsec, 201
Partial pressure, 44
Pascal, 32
Pascal's principle, 35
Pauli exclusion principle, 112
Pedalfer, 249
Pedocal, 249
Pendulum, 27
Penumbra, 182
Period, 116
   geologic, 280
Periodic table, *front cover*, 114
Permanent magnet, 57
Permeability, 270
Pesticides, 284
Ph, 137
Phase, 70, 75, 76, 80, 108
Phase of moon, 181
Phosphates, 284
Phosphor, 164
Photoelectric effect, 92
Photon, 92
Photosphere, 179
Pi bond, 151
Planck's constant, 92

INDEX

Planets, 193
  Jovian, 195
  terrestrial, 194
Plate boundaries, 273, 274
  tectonics, 273–277
Pluto, 196
Point source, 72
Polar:
  easterlies, 229
  front, 229
  ice caps, 194
Polarization, 81
Polyatomic ion, 122
Polymer, 153
Polysaccharides, 154
Potential difference, 52, 56
  energy, 25
Power, 24
Power dissipation, 54
Precambrian time, 280
Precession, 193
Precipitation, 234
Pressure, 32–38, 215–216, 220–224
Pressure gradient force, 220
Prevailing westerlies, 229
Primeval fireball, 207
Principal axis, 83
Principle of equivalence, 98
Prism, 83
Product (chemical), 125
Proportionality, 302
Protein, 156
Proton, 104, 106
Protoplanet, 208
Pulsar, 203
Pyrex glass, 43
Pyroclastics, 264

Quanta, 92
Quantization of energy, 92, 107
Quantum number, 108
Quasar, 206

Radian, 13
Radiation:
  budget, 218
  detectors, 163–164
  of heat, 43
Radioactive decay, 164
Radioactivity, 162
Rapids, 251
Ray tracing, 83–86
Reactants, 125

Reaction, chemical *(see Chemical reaction)*
  rates, 130, 131
Real image, 83
Red giants, 202
Red shift, 202
Reduction, 140
Reflection, 72
Refraction, 74
Refrigerator, 47
Regolith, 188
Relative dating, 279
Relative humidity, 212
Relativity, 94
  general, 97
  special, 95
Representative elements, 123
Resistance, 53, 56
Revolution without rotation, 184
Richter scale, 261
Rift valley, 277
Right ascension, 192
RNA, 157
  mRNA, 160
  tRNA, 160
Rocks, 248–249
Rutherford scattering, 104

Salinity, 253
Salt, 137
Salt bridge, 144
Saltation, 250
Saturation, 133
Saturn, 195
Scalar, 298
Scattering, 79, 104
Schroedinger equation, 109
Scientific notation, 291
Scintillation counter, 164
Sea breeze, 228
Sea floor spreading, 274
Seamount, 253
Seasons, 176
Sea water, 253
Sedimentary rocks, 249
Sedimentation, 285
Seismograph, 263
Seismology, 261–263
Series circuit, 55
Shadow zone, 263
Shear lines, 243
Shells, 114
  filled vs. unfilled, 116–117

Shooting stars, 198
Short range force, 165
SI system, 291
Sidereal day, 173
  month, 182
Significant figures, 293
Sill, 266
Sleet, 236
Slump, 259
Smog, 282
Snow, 236
Soft water, 286
Solar:
  day, 173
  eclipse, 182
  flares, 180
  prominences, 180
  wind, 179
Solenoid, 58
Solid, 30
Solifluction, 259
Solstice, 176
Solubility, 133
Solute, 132
  effect, 235
Solution, 132
Solvent, 132
Soot, 282
Sound, 71
Specific gravity, 248
Specific heat, 40–42
Spectator ion, 134
Spectral class, 200
Spectrum, 78, 90, 107
Speed, angular, 15
  average, 4
  instantaneous, 5
Spring, 270
Stack, 256
Stalactite, 271
Stalagmite, 271
Starch, 154
Star, 199
  cluster, 203
Static electricity, 50
Stellar evolution, 203–204
  parallax, 200
Stratopause, 214
Stratosphere, 214
Streak, 248
Stream, 250
Stream valley, 251
Stress, 260

Striations, 256
Structural formula, 148
Subduction, 273
Sublimation, 31
Substitution reaction, 153
Sucrose, 154
Sulfur dioxide, 282
Sun, 179–180
Sunspots, 180
Superposition principle, 279
Supersaturation, 133, 212
Swamp, 270
Syncline, 267
Synodic month, 182

Telescopes, 86–89
Temperature, 39
  atmospheric, 213, 215, 219
  conversions, 304
  inversion, 284
Thermal equilibrium, 40
Thermally driven circulation, 225
Thermodynamics, 47–49
  first law, 47
  second law, 49
Thermometer, 40
Thermosphere, 214
Thermostat, 43
Thunderstorms, 241–242
Tides, 183–186
Time dilation, 96
Time zone, 174
Tornado, 242
Torque, 17
Trade winds, 228
Transformer, 64
Transition metals, 123
Tropopause, 213
Troposphere, 213
Trough, 70, 230
  glacial, 256

Ultraviolet catastrophe, 92
Umbra, 182
Uncertainty principle, 93
Unsaturated solution, 133
Upper air wave, 230
Upwelling, 254
Uranus, 196

Valence, 121
Vapor pressure, 46
Variable star, 202

INDEX

Vectors, 298–301
Vein, 266
Velocity, 5, 15, 17
Vents, 264
Venus, 194
Vertical ray, 176
Virtual image, 86
Vitamins, 155
Volcanoes, 264–266
Volt, 52
Voltaic cell, 144

Water table, 270
Water treatment, 285–286
Wave, 67
   earthquake, 261–263
   electromagnetic, 72
   function, 109
   longitudinal, 69
   sound, 71
   standing, 69
   transverse, 67
   water, 71
Wave-cut cliffs, 256
Wave-cut platform, 256
Wave-particle duality, 93
Wavefront, 72
Wavelength, 70
Weather, 211
Weathering, 258–259
Weight, 11
Wells, 271
White light, 78
Wind, 220, 223–225, 257
Work, 23

Young's experiment, 79–81

Zodiac, 192
Zonal flow, 228